T0215463

Communications in Computer and Information Science 545

Commenced Publication in 2007
Founding and Former Series Editors:
Alfredo Cuzzocrea, Dominik Ślęzak, and Xiaokang Yang

More information about this series at http://www.springer.com/series/7899

Michael W. Berry · Azlinah Hj. Mohamed
Yap Bee Wah (Eds.)

Soft Computing in Data Science

First International Conference, SCDS 2015
Putrajaya, Malaysia, September 2–3, 2015
Proceedings

 Springer

Editors
Michael W. Berry
University of Tennessee
Knoxville, Tennessee
USA

Yap Bee Wah
Universiti Teknologi MARA
Shah Alam
Malaysia

Azlinah Hj. Mohamed
Universiti Teknologi MARA
Shah Alam
Malaysia

ISSN 1865-0929   ISSN 1865-0937   '(electronic)
Communications in Computer and Information Science
ISBN 978-981-287-935-6   ISBN 978-981-287-936-3   (eBook)
DOI 10.1007/978-981-287-936-3

Library of Congress Control Number: 2015949465

Springer Singapore Heidelberg New York Dordrecht London

Printed on acid-free paper

Springer Science+Business Media Singapore Pte Ltd. is part of Springer Science+Business Media
(www.springer.com)

Preface

We are pleased to present the proceeding of the International Conference on Soft Computing in Data Science 2015 (SCDS 2015). SCDS 2015 was held in the Pullman Hotel, Putrajaya Malaysia, during September 2–3, 2015. The theme of the conference was "Science in Analytics: Harnessing Data and Simplifying Solutions." SCDS 2015 aimed to provide a platform for academics and practitioners to share their knowledge and experience in the area of soft computing in data science. Research collaborations between academia and industry can lead to the advancement of useful analytics and computing applications for providing real-time insights and solutions.

We were delighted that the conference attracted submissions from a diverse group of national and international researchers. We received 69 submissions, among which 25 were accepted. SCDS 2015 utilized a double-blind review procedure. All accepted submissions were assigned to at least three independent reviewers (at least two international reviewers) in order to have a rigorous and convincing evaluation process. A total of 55 international and 79 local reviewers were involved in the review process. The reviewers provided very good feedback and this helped in the selection of papers. The conference proceedings volume editors and Springer's CCIS Editorial Board decided on the final selection of papers. Finally, 25 of the 69 submisssions (36%) were accepted for the conference proceedings.

We would like to thank all researchers who submitted manuscripts to SCDS 2015. We thank the reviewers for volunteering to review the papers. We thank all conference committee members for their time, ideas, and efforts in organizing SCDS 2015. We also wish to thank the Springer CCIS Editorial Board, organizations, and sponsors that supported SCDS 2015.

We sincerely hope that SCDS 2015 provided a venue for sharing knowledge and publishing good research findings and new research collaborations. Last but not least, we hope everyone had an enjoyable and memorable experience at SCDS 2015 in Malaysia.

September 2015

Michael W. Berry
Azlinah Hj. Mohamed
Yap Bee Wah

Organization

Patron

Tan Sri Dato' Sri Prof Ir Dr. Sahol Hamid Abu Bakar, FASc
Vice Chancellor, Universiti Teknologi MARA Malaysia

Honorary Chairs

Michael W. Berry	University of Tennessee, USA
Fazel Famili	University of Ottawa, Canada
Min Chen	Oxford University, UK
Lofti A. Zadeh	University of California, Berkeley, USA
Haibo He	University of Rhode Island, USA
Do van Thanh	Norwegian University of Science and Technology, Norway

Conference Chairs

Azlinah Hj. Mohamed	Universiti Teknologi MARA
Yap Bee Wah	Universiti Teknologi MARA

Program Committee

Rokiah Embong	Universiti Teknologi MARA
Zainab Abu Bakar	Universiti Teknologi MARA
Wan Abdul Rahim Wan Mohd Isa	Universiti Teknologi MARA
Siti Zaleha Zainal Abidin	Universiti Teknologi MARA
Sharifah Aliman	Universiti Teknologi MARA
Shuzlina Abdul Rahman	Universiti Teknologi MARA
Norjansalika Janom	Universiti Teknologi MARA
Mazani Manaf	Universiti Teknologi MARA
Anitawati Mohd Lokman	Universiti Teknologi MARA
Nurazzah Abd Rahman	Universiti Teknologi MARA
Wan Adilah Wan Adnan	Universiti Teknologi MARA
Norehan Abd Manaf	Universiti Teknologi MARA
Norizan Mat Diah	Universiti Teknologi MARA
Shuhaida Shuhidan	Universiti Teknologi MARA

Student Committee

Norulhidayah Isa	Universiti Teknologi MARA
Hezlin Aryani Abd Rahman	Universiti Teknologi MARA
Hamzah Bin Abdul Hamid	Universiti Teknologi MARA
Mohamad Arsyad Yakob	Universiti Teknologi MARA

International Scientific Committee

Simon Fong	University of Macau, SAR China
Mohammed Bennamoun	University of Western Australia
Yasue Mitsukura	Keio University, Japan
Dhiya Al-Jumeily	Liverpool John Moores University, UK
Dariusz Krol	University of Wroclaw, Poland
Richard Weber	University of Chile, Santiago, Chile
Jose Maria Pena	Technical University of Madrid, Spain
Ali Selamat	Universiti Teknologi Malaysia
Daud Mohamed	Universiti Teknologi MARA
Yusuke Nojima	Osaka Perfecture University, Japan
Norita Md Nawawi	Universiti Sains Islam Malaysia
Siddhivinayak Kulkarni e	Universiti of Ballarat, Australia
Tahir Ahmad	Universiti Teknologi Malaysia
Saeed Aghabozorgi	Universiti Malaya, Malaysia
Shanudin Zakaria	Universiti Pertahanan Malaysia
Ku Ruhana Ku Mahamud	Universiti Utara Malaysia
Azuraliza Abu Bakar	Universiti Kebangsaan Malaysia
Xian-Jin Xie	University of Texas Southwestern Medical Center, USA

International Reviewers

Abdullah Al Luhaib	KSU, Saudi Arabia
Akhil Jabbar	Aurora's Engineering College, Bhongir, India
Alaa Aljanaby	University of Nizwa, Oman
Alexandre Torrezam	Science and Technologie of Mato Grosso - IFMT, Brazil
Ali Naderi Saatlo	Istanbul Technical University, Turkey
Aryo Pinandito	Universitas Brawijaya, Indonesia
Bashar Al-Shboul	The University of Jordan
Bassam Al-Jewad	Advanced Technology Systems-Iraq LLC
Davide Anguita	University of Genova, Italy
Devendra Chaturvedi	Deemed University, India
Fazel Famili	University of Ottawa, Canada
Gholamreza Akbarizadeh	Shahid Chamran University, Iran
Haibo He	University of Rhode Island, USA
Haider M. AlSabbagh	Basra University, Iraq

Local Reviewers

Abdul Rahman Othman	Universiti Sains Malaysia
Abdullah Embong	University Malaysia Pahang
Adib Kabir Chowdhury	University College of Technology Sarawak
Ahmad Rafi	Multimedia University
Ahmad Shahrizan Abdul Ghani	University Sains Malaysia
Ahmad Sobri Hashim	Universiti Teknologi PETRONAS
Alfian Abdul Halin	Universiti Putra Malaysia
Ali Qusay Al-Faris	Universiti Sains Malaysia
Ali Selamat	Universiti Teknologi Malaysia
Arsmah Ibrahim	Universiti Teknologi MARA
Asmala Ahmad	Universiti Teknikal Malaysia Melaka
Asmidar Abu Bakar	Universiti Tenaga Nasional
Azilawati Azizan	Universiti Teknologi MARA
Azlinah Mohamed	Universiti Teknologi MARA
Azuraliza Abu Bakar	Universiti Kebangsaan Malaysia
Bahbibi Rahmatullah	Universiti Pendidikan Sultan Idris
Bhagwan Das	Universiti Tun Hussein Onn Malaysia
Bong Chih How	Universiti Malaysia Sarawak
Choong-Yeun Liong	Universiti Kebangsaan Malaysia
Fatimah Dato Ahmad	Universiti Pertahanan Nasional Malaysia
Haidy Henry	Universiti Teknologi MARA
Hamidah Jantan	Universiti Teknologi MARA
Hasimah Mohamed	Universiti Sains Malaysia
Haslizatul Mohamed Hanum	Universiti Teknologi MARA
Hasniza Yahya	Universiti Tenaga Nasional
K. Rao	Birla Institute of Technology and Science, Malaysia
Kamaruddin Mamat	Universiti Teknologi MARA
Kamarul Abdul-Basit	Universiti Teknologi MARA
Kamarularifin Abd Jalil	Universiti Teknologi MARA
Kwang Hooi Yew	Universiti Teknologi PETRONAS
Lilly Suriani Affendey	Universiti Putra Malaysia
Maizura Mohd Sani	Universiti Teknologi MARA
Marina Yusoff	Universiti Teknologi MARA
Mas Rina Mustaffa	Universiti Putra Malaysia
Mazani Manaf	Universiti Teknologi MARA
Mohamad Daud	Universiti Teknologi MARA
Mohamad Faizal Baharom	Universiti Teknikal Malaysia
Mohamed Imran Mohamed Ariff	Universiti Teknologi MARA

Mohd Alias Lazim	Universiti Teknologi MARA
Mohd Fadzli Marhusin	Universiti Sains Islam Malaysia
Mohd Helmy Abd Wahab	Universiti Tun Hussein Onn Malaysia
Mohd Shamrie Sainin	Universiti Utara Malaysia
Mohd Tahir Ismail	Universiti Sains Malaysia
Muhamad Taufik Abdullah	Universiti Putra Malaysia
Muhammad Rozi Malim	Universiti Teknologi MARA
Mustafa Man	University Malaysia Terengganu
Nasiroh Omar	Universiti Teknologi MARA
Noor Azilah Muda	Universiti Teknikal Malaysia
Noor Elaiza Abd Khalid	Universiti Teknologi MARA
Nordin Abu Bakar	Universiti Teknologi MARA
Noreen Arshad	Universiti Teknologi PETRONAS
Norihan Abdul Hamid	Universiti Teknikal Malaysia
Nur Atiqah Sia Abdullah	Universiti Teknologi MARA
Nur Fadilah Ab Aziz	Universiti Tenaga Nasional
Rafikha Aliana A. Raof	Universiti Malaysia Perlis
Rizauddin Saian	Universiti Teknologi MARA
Rosalina Abd. Salam	Universiti Sains Islam Malaysia
Saidah Saad	Universiti Kebangsaan Malaysia
Shahrul Azman Noah	Universiti Kebangsaan Malaysia
Shahrul Badariah Mat Sah	Universiti Teknikal Malaysia
Sharifalillah Nordin	Universiti Teknologi MARA
Shuzlina Abdul-Rahman	Universiti Teknologi MARA
Siti Zaleha Zainal Abidin	Universiti Teknologi MARA
Sofianita Mutalib	Universiti Teknologi MARA
Sumarni Abu Bakar	Universiti Teknologi MARA
Suzana Ahmad	Universiti Teknologi MARA
Suziah Sulaiman	Universiti Teknologi PETRONAS
Syahrul Nizam Junaini	Universiti Malaysia Sarawak
Tahir Ahmad	Universiti Teknologi Malaysia
Waidah Ismail	Universiti Sains Islam Malaysia
Weng Kin Lai	Tunku Abdul Rahman University College
Yap Bee Wah	Universiti Teknologi MARA
Yuhanis Yusof	Universiti Utara Malaysia
Yun Huoy Choo	Universiti Teknikal Malaysia
Zaidah Ibrahim	Universiti Teknologi MARA
Zainab Abu Bakar	Universiti Teknologi MARA
Zolidah Kasiran	Universiti Teknologi MARA
Zuhaimy Ismail	Universiti Teknologi Malaysia

Contents

Data Mining

Fuzzy Computing

Evolutionary Computing/Optimization

Pattern Recognition

Human Machine Interface

Hybrid Methods

Part I
Data Mining

An Improved Particle Swarm Optimization via Velocity-Based Reinitialization for Feature Selection

Shuzlina Abdul-Rahman[1], Azuraliza Abu Bakar[2], and Zeti-Azura Mohamed-Hussein[3]

[1] Faculty of Computer and Mathematical Sciences,
Universiti Teknologi MARA, Shah Alam, Selangor, Malaysia
[2] Faculty of Information Science and Technology, UKM, Malaysia
[3] Faculty of Science and Technologi, UKM, Malaysia
shuzlina@tmsk.uitm.edu.my,
{azuraliza,zeti}@ukm.my

Abstract. The performance of feature selection method is typically measured based on the accuracy and the number of selected features. The use of particle swarm optimization (PSO) as the feature selection method was found to be competitive than its optimization counterpart. However, the standard PSO algorithm suffers from premature convergence, a condition whereby PSO tends to get trapped in a local optimum that prevents it from being converged to a better position. This paper attempts to improve the velocity-based initialization (VBR) method on the feature selection problem using support vector machine classifier following the wrapper method strategy. Five benchmark datasets were used to implement the method. The results were analyzed based on classifier performance and the selected number of features. It was found that on average, the accuracy of the particle swarm optimization with an improved velocity-based initialization method is higher than the existing VBR method and generally generates a lesser number of features.

Keywords: Feature Selection, Particle Swarm Optimization, Velocity-based Reinitialization.

1   Introduction

Particle swarm optimization (PSO) is an optimization technique in the field of machine learning which provides an evolutionary based search introduced by Kennedy & Eberhart [1]. Similar to ant colony optimization (ACO) that attempts to mimic the behaviour of ants, PSO was inspired by the behaviour of flocks of birds and shoals of fish. Since then many versions of PSO have been proposed in past studies [2-5]. PSO algorithms are especially useful for parameter optimization in continuous and multi-dimensional search spaces. PSO has shown its success to a variety of problems, including feature selection [6-10]. The principal concept of PSO deals with the creation of a swarm of candidate solutions in which each potential solution is seen as a particle with a particular rate of change or velocity that operates through the search space. Each particle retains its own individual memory or the best position it has visited, as

© Springer Science+Business Media Singapore 2015
M.W. Berry et al. (Eds.): SCDS 2015, CCIS 545, pp. 3–12, 2015.
DOI: 10.1007/978-981-287-936-3_1

well as a global memory of the best position visited by all particles in the swarm so far. The velocity and position are updated according to the following Equation 1 and Equation 2.

$$v_{id} = w.v_{id} + \varphi_1.U(0,1).(p_{id} - x_{id}) + \varphi_2.U(0,1).(p_{gd} - x_{id}) \tag{1}$$

$$x_{id} = x_{id} + v_{id} , d = 1,2,..., S \tag{2}$$

Where w is the inertia weight; φ_1 and φ_2 are the acceleration positive constants known as cognitive learning rate and social learning rate respectively; $U(0,1)$ is a random function within the range [0,1]. The velocity update in Equation 1 contains three essential parameters for PSO: the momentum component, the cognitive component and the social component. The first component guides how much the particle recalls its previous velocity via its inertial constant, w. The second component guides how much the particle heads towards its personal best via its cognition learning factor, φ_1. The last component attracts the particle towards swarm's best ever position via its social learning factor, φ_2. Using Equation 1, the new velocity of the particle will be calculated based on its previous velocity as well as the distances of its current position from its own best experience and the group's best experience [8]. Subsequently, the particle flies toward a new position according to Equation 2. In the feature selection (FS) context, each solution of the particle is represented as fixed length binary strings (i.e. $X_i = (x_{i1}, x_{i2},....., x_{iN})$, in N dimensional search space, where $x_{id} \in \{0,1\}$, $i = 1, 2, ..., n$ and $d = 1, 2, ..., N$). The coordinates x_{id} of these particles have a velocity, $v_{i} = (v_{i1}, v_{i2}, ..., v_{iN})$, where $v_{id} \in \{0,1\}$, $i = 1, 2, ..., n$ and $d = 1, 2,..., N$.

Each particle has its own fitness value that needs to be optimised. In FS problem, evaluation of the performance or fitness function of any particles i is usually computed based on the classification accuracy and the selected number of features. The formula for the fitness function is then defined as shown in Equation 3 [8]. The two parameters, α and β determine the significance of these two principles. $\gamma_{Fsel}(D)$ is the classification rate of condition feature set F relative to decision D, F_{all} is the initial number of features, and F_{sel} is the selected features. The pseudo-code of the PSO algorithm can be found in Wang et al. [8].

$$fitness = \alpha^*\gamma_{F_{sel}}(D) + \beta^* \frac{|F_{all}| - |F_{sel}|}{|F_{all}|} \tag{3}$$

The key contributions of this work are the proposal of the particle swarm optimization with an improved velocity-based initialization (PSO_ImVBR). The PSO_ImVBR is compared with the existing particle swarm optimization with velocity-based initialization (PSO_VBR) where the support vector machine (SVM) algorithm acts as a classifier. The use of PSO is motivated by two factors. First, compared to genetic algorithm, the operation of PSO does not involve crossover and mutation, thus it is computationally inexpensive, both memory and runtime [8]. Second, unlike other heuristic techniques, PSO has a flexible and well-balanced mechanism to enhance the global and local exploration abilities [11]. The remainder of this paper is organised as follows. Section 2 discusses some related works related to the velocity-based reinitialization (VBR) approaches. Section 3 presents the proposed method. In Section 4, the experimental setup is presented and followed by the results discussion in section 5. Finally, in Section 6, the conclusion is given.

2   Related Works

This section reviews past studies that work around the VBR methods and some other improvement of PSO algorithms. Although PSO is superior to its counterparts such as GA, it still has some weaknesses and one common problem is the premature convergence; a condition whereby the global best particle gets stuck in a local optimum. The remaining particles will also converge to the same position and their updated velocities also approach zero. Finally, they lose exploration capability and stop moving. This condition would prevent them from being converged towards a better position or a global optima solution [12-13]. Various modified techniques have been proposed to tackle this problem by many means [5, 9, 12, 14-17].

There are several ways to avoid the premature convergence. One way is by using some form of mutation operators similar to GA [5, 18-19]. In the work of Higashi & Iba [19], a mutation operation refines the values of a particle position based on a random number resulting from a Gaussian distribution. In contrast to Higashi & Iba [19] that focused on particle position, Ratnaweera et al. [5] executed the mutation operator on a particle velocity if the global best position remained unchanged for certain predetermined iterations. The mutation was applied on PSO mutation with the time-varying inertia weight (MPSO-TVIW) and resulted in a significant improvement. However, the mutation step size and mutation probability were less sensitive for most of the tested functions. In another study by Wang et al. [18], a dynamic cauchy mutation was applied on the global best particle in every generation. The method was inspired by the opposition-based learning method introduced by Tizhoosh [20]. The method labelled as opposition-based particle swarm optimization (OPSO) was compared to the existing PSO, and showed competitive results. However, there are still cases where the method may not able to prevent the search from falling in the local optimal condition which demands more study.

An alternative method to avoid the premature convergence is by reinitialization, either reinitialising the global best or the velocity. This kind of method is mentioned in many past studies [5, 9, 12, 21-22]. Both Chuang et al. [9] and Yang et al. [22] monitor the premature convergence on global best. If the global best values carry the same value for a predetermined generation, the global best position is reset to zero which means no features are selected [9]. In the work of Yang et al. [22], a further step is taken whereby a simple boolean algebra operation was applied on pbest of all particles. The logical operator 'and' was used to 'and' pbest of all the particles to form a new binary string that is considered as a new global best. The method by Huang et al. [14] however employs a shallow Boolean algebra operation that does not show much improvement in accuracy and was only tested on a single domain (i.e. microarray data). Another variant of the velocity reinitialization was proposed in Binkley & Hagiwara [12] namely velocity-based initialization (VBR). Their study proposed a new scheme for identifying the demand of complete swarm initialization. The algorithm operates similarly to the existing PSO with an additional premature convergence or stagnation function that evaluates when most of the swarms are not being used effectively. The novelty in Binkley & Hagiwara's [12] work is the use of the median-based method for reinitialization. A more reliable method on velocity reinitialization was proposed in the work of Worasucheep [17]. The method was known as particle swarm optimization with stagnation detection and dispersion (PSO-DD) exploits both global fitness and velocity parameter to identify the stagnant condition. The PSO_ImVBR proposed in this study was inspired by the work of Binkley & Hagiwara [12] and Worasucheep [17]. The proposed method integrates these two methods that worked on the basis of velocity reinitialization.

The researcher's method differs than the earlier works in several ways. First, the PSO_ImVBR was an attempt that investigated the VBR in the feature selection problem involving datasets of high dimensionality. With an exception of Chuang et al. [9] and Yang et al. [22] that limits feature selection problems on microarray datasets, all past studies that work around the velocity reinitialization focuses on multimodal function optimization [5, 12, 17, 21]. Second, the work of Binkley & Hagiwara [12] suffers two limitations. The inspection of the stagnant was done at every iteration which added computational complexity to the algorithm, while the value of threshold was difficult to determine. The suggested value of the threshold was inappropriate to be implemented in feature selection problems. The researcher's preliminary experiments showed that those threshold values did not benefit the feature selection problem. Thus, another method to determine the threshold values had to be considered. However the idea of velocity median that they employed seems worth implementing and was embedded in the researcher's work. Third, in contrast to Worasucheep's [17] work that reset half way the inertia weight back to its original value, the researcher's work used the PSO-variants with a constriction coefficient parameter without the need of resetting the weights, thus simplifying the overall processes. However, Worashucheep's [17] work that employed both the velocity and global fitness gave some merits. These two parameters provided good indicators as when to reinitialise the swarm.

3   The Proposed PSO_ImVBR Method

The PSO_ImVBR method contained an improvement ratio that integrated velocity and global fitness parameters, two important parameters in PSO algorithms as adopted in Worasucheep [17]. These two parameters were used to monitor the stagnant condition. The improvement ratio (IR) formula monitored an improvement of global fitness best over an improvement of velocity of all the particles in the swarm. For the maximization problem, the global fitness was expected to increase whereas the velocity was expected to decrease over the total iterations. The formula of IR is adapted from Worasucheep [17] and shown in Equation 4.

$$\text{Improvement Ratio} = IR = \frac{\left|1-\frac{gf_k}{gf_{k-1}}\right|}{\left|1-\frac{mv_k}{mv_{k-1}}\right|} \tag{4}$$

Where; gf_k is the global fitness at iteration k; gf_{k-1} is the global fitness at iteration k-1; mv_k is the median velocity at iteration k of the swarm; mv_{k-1} is the median velocity at iteration k-1 of the swarm.

In contrast to Worasucheep [17] that used the average velocity, the researcher made use of the median velocity that was adopted in Binkley & Hagiwara's [12] work. The median velocity was adopted in order to confirm that a swarm is stagnant when the majority of the particles have stagnated. There exist two conditions during the calculation of the median of the velocity vector. If the number of total velocities is odd, the median of the velocity is the middle value otherwise the median of the velocity is the average of the middle two values. During an optimization run, the particles tend to move slower with their median velocity approaching zero. Besides, the global best is also hardly improved and the value of gf_k / gf_{k-1} is approaching one. If this condition occurs, the IR ratio has dropped below a certain stagnation threshold and the stagnation is expected. Subsequently, the current swarm's global best is kept in the swarm global best list for later use and the swarm is completely reinitialised. In doing this operation, a linked-list is created to store all the relevant values. Every time the stagnant condition is identified, the values such as global best, the accuracy, the features dimension (length) and modelling time are kept in the linked-list. The initialization means that both the velocity vectors and the particle's position are randomly initialised. Once the predetermined stopping criterion is satisfied, a loop is triggered to check the linked-list and return the highest parameter value. The first loop checks the accuracy and if there exist several similar accuracies, the second loop is performed to check the features dimension and the one with the lowest feature dimension is selected. Similarly, if there exist several similar features dimension, the third loop is performed to get the lowest modelling time.

The algorithm begins the process similar to the existing PSO algorithm with an additional stagnation function that has to be performed prior to reinitialising the swarm. As opposed to the VBR method proposed in Binkley & Hagiwara [12] where the premature convergence check is performed in every iteration, the checking was performed only after the first fifth of the total generations and continued the checking after a certain elapse generation. After some preliminary experiments, the researcher found that most datasets become stagnant after the first fifth of generations. For example, if there is 100 generations and the elapse generation is 30, the reinitialization will begin on the 50th generation (i.e. $\left(\frac{1}{5}x100\right) + 30 = 50$). The subsequent reinitialization will be at 80th generation and so forth. Therefore, the computational complexity was reduced besides allowing the swarm to explore the search space within the predetermined boundary. Theoretically, during an optimization run, the median velocity of the particles rapidly approach zero. To ensure an efficient initialization without sacrificing the previous global best value, the best twenty-percent of particles is retained from being initialised. The other eighty-percent of particles is initialised in which the global best is set to zero and the particle's position and velocity is initialised accordingly. The reason is to keep the potential solutions for the next itera-tion and to avoid losing promising regions. The elapse generation and percentage were chosen based on our preliminary exercise on the algorithm using several data-sets. The algorithm 1 is adapted from Binkley & Hagiwara [12] and the algorithm 2 is the improved algorithm that details the PSO_ImVBR method.

Algorithm 1: The PSO_ImVBR

1.	Create and initialise population
2.	Set global best list to empty
3.	While stopping condition not reached do
4.	stagnant() //refer to Algorithm 2
5.	For i = 1 to population size do
6.	For d =1 to number of dimensions do
7.	r_1=rand() //between 0 and 1
8.	r_2=rand() //between 0 and 1
9.	Update velocities
10.	Update positions
11.	End for
12.	If particle i's position is better than its personal best then update particle i's personal best
13.	End if
14.	Endfor
15.	Update the swarm global best with the best of all particle personal bests
16.	End while
17.	Add swarm's global best to global best list
18.	Return the best of the global best list

Algorithm 2: Identifying stagnation in PSO_ImVBR

1.	If generation is one fifth of total generations then
2.	For every x generation
3.	Get the euclidean norm of each velocity vector and sort the norms
4.	Get the velocity$_{medianCurr}$ and velocity$_{medianPrev}$
5.	Get the global_fitness$_{current}$ and global_fitness$_{previous}$
6.	Calculate the improvement ratio using Equation 4
7.	If improvement ratio < threshold then
8.	Add global$_{best}$ to global best list
9.	Reinitialise the swarm
10.	End if
11.	End for
12.	End if

---

4   Experimental Setup

This section presents the experimental setup of the study. The PSO algorithm was implemented in NetBeans IDE 6.9 in a PC with the following features: Intel Core 2 Duo, 2.2 GHz CPU, 2G RAM, a Windows Vista operating system. Prior to optimizing the dataset with the PSO algorithm, the features were filtered using the multivariate filter namely the correlation feature selection with the linear forward search (CFS-LFS) approach. The costly computational complexity of PSO algorithm suggests the features dimension to be pre-reduced by the filter approach. Table 1 presents the profile of the datasets used in this study with a minimum of 168 features and maximum of 649 features. The selected datasets are commonly used in data mining and machine learning research [23-25]. The instances of the Musk2 dataset was reduced from the original instances using the stratified remove folds and labeled as [red]. In this study, a preliminary experiment was conducted to determine the most suitable PSO'S parameter and was settled with 30 particles and 100 generations.

Table 1. Feature summary of dataset

Description	Domain	Feature Type	Size	Instances	Classes
Musk2 [red]	Chemistry	Nominal	168	2200	2
Promoters	DNA data	Numeric	228	106	2
Splice	DNA data	Numeric	240	3190	3
Arrhythmia	ECG data	Categorical, Numeric	279	452	16
Multifeatures	Mixed	Numeric	649	2000	10

Nom: Nominal; Num: Numeric

The first compared method was PSO_VBR and the second compared method was PSO_Yang [22]. This method identifies the need of swarm reinitialization by monitoring the velocities throughout the iterations. If the median of the velocity has dropped below a certain threshold value, the whole swarm is initialized. The stagnation modules in the PSO_VBR only check the velocity median with the threshold value whereas the researcher's method combined both the velocity median and the fitness function values. Binkley & Hagiwara [12] employed three threshold values, $\alpha = 0.01$, $\alpha = 0.001$, and $\alpha = 0.0001$. However, during implementation the researcher found difficulty in setting the threshold value. These values, perhaps suited to his problems but did not benefit the researcher's problem that addressed the features dimension. Thus, in this study, the researcher settled the value of α by half of the total features dimension after exercising several different values of α on two-third of the datasets. Thus, different dataset uses different values of α. The velocity in the researcher's method was randomly initialized and constrained by the maximum allowed velocity [-6, +6].

5   Results and Discussion

The experimental results for the five datasets are summarized in Table 2 and Table 3. The last two rows summarize the average results of each criteria and the wins/ties/losses (at p-value <0.05) over the PSO_VBR method. The same definition of the good trend applies in this discussion whereby the classification accuracy is maintained (given by the number of ties) or increased (given by the number of wins) despite the dimensionality reduction. Measuring classification accuracy, Table 2 records the results of the experiments for each of the compared method against the PSO_VBR method. There are several observations which can be drawn from these results. First, both of the comparison method achieved higher accuracy than the PSO_VBR method. It was found that on average, the accuracy of the PSO_ImVBR is higher than PSO_Yang method. In terms of t-test, it was found that the PSO_ImVBR yielded more number of wins compared to the PSO_Yang method with three significant wins and two ties. The improvement velocity that was embedded in the PSO_ImVBR method helps to improve the accuracy of the classifier. Next, the researcher further analyzed the selected features generated from these three methods.

Table 2. Comparison of classification accuracy

Datasets	PSO_VBR (1)	PSO_ImVBR (2)	PSO_Yang (3)	p_val_1 (1-2)	p_val_2 (1-3)
Musk2 [red]	93.9	94.2^W	93.7^T	0.003+	1.000
Promoters	100.0	100.0^T	100.0^T	1.000	1.000
Splice	94.7	95.4^W	94.5^T	0.000+	1.000
Arrhythmia	80.8	81.1^T	80.3^T	0.146	1.000
Multifeatures	99.2	99.4^W	98.9^T	0.021+	1.000
Avg/WTL	93.6	94.0 [3/2/0]	93.5 [0/5/0]		

Table 3 records the average selected features from the three methods. In this criterion, the researcher aimed to investigate whether the selected features could be further reduced with the improved VBR. On average all the three methods achieved a significant reduction of dimensionality by selecting only a small portion of the features from the multivariate (MV) Filter phase. There was about 55-56% of the features selected from the MV Filter phase. On average, all the three compared method have selected competitively similar number of features. Inspecting closely, the PSO_ImVBR method has selected a lesser number of features in three out of five datasets compared to PSO_VBR (Musk2, Promoters, Arrhythmia, Multifeatures) and four out of five datasets compared to PSO_Yang (Musk2, Arrhythmia, Multifeatures). Thus the researcher concludes that the PSO_ImVBR tends to generate a lesser number of features if compared to the other two methods. The reason could be due to the more exploration exposure opportunities that the particles found during the initialization process. The VBR gave more chances to the particles to explore new areas from being trapped in a local optimum solution.

The optimization solution given by the PSO_ImVBR was considered successful from the following trends. The PSO_ImVBR method improves or maintains the accuracy of the classifier in majority of the datasets. In general, the PSO_ImVBR method achieved a significantly higher number of wins compared to PSO_VBR and PSO_Yang. These findings showed that the improved VBR does assist in selecting fewer features while maintaining or improving the accuracy. The good trends were able to be accomplished due to the mechanism of VBR that allows the particles to explore to the other new areas that perhaps promised a better solution and avoid from being trapped in a local optimum solution. Besides, the initialization formula that retains the best twenty percent of particles from being initialized also helps to ensure the efficiency of the method.

Table 3. Average (Avg.) selected features (SF) by each PSO method

Datasets	#Features	PSO_VBR	PSO_ImVBR	PSO_Yang
	(CFS-LFS)	Avg. SF (%)	Avg. SF (%)	Avg. SF (%)
Musk2 [red]	22	14.7 (66.8)	14.2 (64.5)	14.5 (65.9)
Promoters	8	3.0 (37.5)	3.0 (37.5)	3.0 (41.8)
Splice	27	18.5 (68.5)	19.8 (73.3)	19.3 (69.7)
Arrhythmia	19	12.0 (63.2)	10.5 (55.3)	10.5 (64.3)
Multifeatures	134	63.9 (47.7)	62.0 (46.3)	62.0 (47.1)
Average	32	22.4 (56.7)	21.9 (55.4)	21.9 (55.3)

6   Conclusion

The study proposed the use of the VBR method in the PSO to overcome the premature convergence that hinders PSO algorithm. The findings from this study revealed that the PSO_ImVBR was generally significantly better than the exiting PSO_VBR and PSO_Yang methods. The PSO_ImVBR method was able to give the highest accuracy with less selected features. The analysis also showed that the proposed method was able to overcome the premature convergence as compared to the PSO_Yang. Overall, the researcher concluded that the PSO_ImVBR helps the stagnation and is suitable to be applied in the feature selection problem.

Acknowledgment. This project was funded by the Universiti Teknologi MARA, Shah Alam, Selangor, MALAYSIA.

References

1. Kennedy, J., Eberhart, R.C.: Particle swarm optimization. In: Proceedings of IEEE International Conference on Neural Networks, Piscataway, NJ, pp. 1942–(1948)
2. Clerc, M., Kennedy, J.: The Particle Swarm - Explosion, Stability, and Convergence in a Multidimensional Complex Space. IEEE Transactions on Evolutionary Computation 6, 58–73 (2002)
3. van den Bergh, F., Engelbrecht, A.P.: Analysis of Particle Swarm Optimizers. Information Sciences 176, 937–971 (2006)
4. Kadirkamanathan, V., et al.: Stability analysis of the particle dynamics in particle swarm optimizer. IEEE Transactions on Evolutionary Computation 10, 245–255 (2006)
5. Ratnaweera, A., et al.: Self-organizing hierarchical particle swarm optimizer with time-varying acceleration coefficients. IEEE Transactions on Evolutionary Computation 8, 240–255 (2004)

6. Pedrycz, W., et al.: Identifying core sets of discriminatory using particle swarm optimization. Expert Systems with Applications 36, 4610–4616 (2009)
7. Meissner, M., et al.: Optimized Particle Swarm Optimization (OPSO) and its application to artificial neural network training. BMC Bioinformatics 7, 125 (2006)
8. Wang, X., et al.: Feature selection based on rough sets and particle swarm optimization. Pattern Recognition Letters 28, 459–471 (2007)
9. Chuang, L.-Y., et al.: Improved binary PSO for feature selection using gene expression data. Computational Biology and Chemistry 32, 29–38 (2008)
10. Wang, L., Yu, J.: Fault feature selection based on modified binary PSO with mutation and its application in chemical process fault diagnosis. In: Wang, L., Chen, K., S. Ong, Y. (eds.) ICNC 2005. LNCS, vol. 3612, pp. 832–840. Springer, Heidelberg (2005)
11. Abido, M.A.: Optimal power flow using tabu search algorithm. Electric Power Components & Systems 30, 469–483 (2002)
12. Binkley, K.J., Hagiwara, M.: Balancing Exploitation and Exploration in Particle Swarm Optimization: Velocity-based Reinitialization, Keio, Japan (2008)
13. Bae, C., et al.: Feature selection with Intelligent Dynamic Swarm and Rough Set. Expert Systems with Applications 37, 7026–7032 (2010)
14. Huang, C.-J., et al.: Application of wrapper approach and composite classifier to the stock trend prediction. Expert Systems with Applications 34, 2870–2878 (2008)
15. Wang, H., et al.: Opposition-based particle swarm algorithm with cauchy mutation. In: IEEE Congres on Evolutionary Computation, pp. 4750–4756 (2007)
16. Ratnaweera, A., et al.: Self-Organizing Hierarchical Particle Swarm Optimizer With Time-Varying Acceleration Coefficients. IEEE Transactions on Evolutionary Computation 8, 240–255 (2004)
17. Worasucheep, C.: A particle swarm optimization with stagnation detection and dispersion. In: IEEE Congress on Evolutionary Computation 2008, Hong Kong, China, pp. 424–429 (2008)
18. Wang, H., et al.: Opposition-based particle swarm algorithm with cauchy mutation. In: Proceedings of the 2007 IEEE Congress on Evolutionary Computation, pp. 4750–4756 (2007)
19. Higashi, N., Iba, H.: Particle swarm optimization with gaussian mutations. In: Proceedings of the IEEE Swarm Intelligence Symposium, pp. 72–79 (2003)
20. Tizhoosh, H.R.: Opposition-based learning: A new schema for machine intelligence. In: International Conference on Computational Intelligence for Modeling Control and Automation (CIMCA 2005), Vienna, Austria (2005)
21. Pasupuleti, S., Battiti, R.: The gregarious particle swarm optimizer (G-PSO). In: Proceedings of the 8th Annual Conference on Genetic and Evolutionary Computation, pp. 67–74 (2006)
22. Yang, C.-S., et al.: A Hybrid Feature Selection Method for Microarracy Classification. IAENG International Journal of Computer Science 35 (2008)
23. Zhu, Z., et al.: Wrapper–Filter Feature Selection Algorithm Using a Memetic Framework. IEEE Transactions on Systems, Man, and Cybernetics-Part B 37, 70–76 (2007)
24. Yu, L., Liu, H.: Efficient Feature Selection via Analysis of Relevance and Redundancy. Machine Learning Research 5, 1205–1224 (2004)
25. Lin, S.-W., et al.: Particle swarm optimization for parameter determination and feature selection of support vector machines. Expert Systems with Applications 35, 1817–1824 (2008)

Classifying Forum Questions Using PCA and Machine Learning for Improving Online CQA

Simon Fong[1], Yan Zhuang[1], Kexing Liu[1], and Shu Zhou[2]

[1] Department of Computer Information Science, University of Macau, Macau SAR
[2] Department of Product Marketing, MOZAT Pte Ltd, Singapore
{ccfong,syz,mb45462}@umac.mo,
suzyzhou@mozat.com

Abstract. As one of the most popular e-Business models, community question answering (CQA) services increasingly gather large amount of knowledge through the voluntary services of the online community across the globe. While most questions in CQA usually receive an answer posted by the peer users, it is found that the number of unanswered or ignored questions soared up high in the past few years. Understanding the factors that contribute to questions being answered as well as questions remain ignored can help the forum users to improve the quality of their questions and increase their chances of getting answers from the forum. In this study, feature selection method called Principal Component Analysis was used to extract the factors or components of the features. Then data mining techniques was used to identify the relevant features that will help predict the quality of questions.

Keywords: Community Question Answering, Principal Component Analysis, Machine Learning, Business Intelligence.

1   Introduction

CQA is defined as community services which allow users to post questions for other users to answer or respond [1]. It aims to provide community-based [2] knowledge creation services [3]. Lately, CQA websites specifically in the programming context are gaining momentum among programmers and software developers [4]. CQA can provide them a forum for seeking help and advice from their professional peers about technical difficulties that they face.

One of the most popular programming CQA website currently, Stack Overflow, managed to capture compelling technical knowledge sharing among software developers globally [5]. Registered members in Stack Overflow can vote on questions and also answers. The positive and negative votes show the helpfulness and quality of a question and answer. There is a reputation system in Stack Overflow, the members can increase their reputation in the website by participating in various activities like posting questions, answering, voting, posting comments, etc. With better reputations, they can be upgraded with extra capabilities such as editing question/answers and closing a topic.

© Springer Science+Business Media Singapore 2015
M.W. Berry et al. (Eds.): SCDS 2015, CCIS 545, pp. 13–22, 2015.
DOI: 10.1007/978-981-287-936-3_2

This study aims to examine the predictors of ignored questions in a CQA service specifically those posted in Stack Overflow, by using machine learning. Thus, there are two main objectives in this study.

The first objective is the identification of the crucial factors or features that affect the quality of the questions. The quality of the questions is divided into two classes: good and bad questions. In this specific context, good questions are defined as the questions that are solved by the community members. Contrarily, bad questions are defined as the ignored questions, which specifically mean the questions without any answers or comments from the online community for at least three months.

The second objective is to investigate the use of feature selection on classification models by using principal component analysis for improving the accuracy of machine learning algorithms, in classifying between good and bad questions. Feature selection technique is used to infer the importance of the features pertaining to the quality of questions in CQA. In this context, an ignored question is defined as question that is without any answers or comments from the community for at least three months.

2   Experiment

The importance of factor analysis, in the context of being a prediction model, is about statistically evaluating how important or significant each model attribute is, pertaining to predicting an outcome from an induced model. In this paper we report an experimentation of a classification model that is built over the historical dataset of a CQA service forum.

The objective of the experimentation is in two-fold. The classification model is potentially being used for testing new question posted to the forum so to estimate its chance of being answered – so called the acceptance rate. This is done by comparing the attributes of the new questions to those that have learned by the model from the historical records of the CQA forum, both that have received replies successfully and otherwise. Feature selection is used in the data pre-processing prior to constructing the model for enhancing the classification accuracy. In a nutshell, feature selection is a computational method that chooses a subset of features or attributes for representing the full feature set. It helps reduce the dimensionality of the classification problem. As a desirable side-effect, feature selection algorithm estimates certain 'contributing factors' for the model attributes. Such contributing factors are perceived as the extent of significance for comparing the relative impacts of the attributes on the predicted outcomes.

2.1   CQA Dataset and Motivation

Stack Overflow's data is used for this study due to the popularity of Stack Overflow among programmers globally. The data are rich in metadata such as user's reputation that are suitable to be used for the study. The topic, Java, is chosen for experimentation in Stack Overflow because this topic has been long-term favored in the past decade. Surely the topic has the highest average ratings in popularity from

2002 to 2014. Java-related questions are the most tagged and also the most ignored, with 456,748 and 8,463 respectively. The data for experimentation are collected from Stack Overflow during October 2014. The Data Explorer service provided by Stack Exchange is applied to obtain the data. SQL queries are be written and executed in Data Explorer to crawl the required data from the database of Stack Overflow. For this experimentation, we crawled the data with the tag 'Java' of solved and ignored questions starting from year 2008 onwards. After that, disproportionate stratified sampling are be used to sample the data from the two categories. A total of 3,000 data are to be sampled, of which 50% are the good questions and another 50% are the bad questions according to the following criterion.

The prediction labels, good questions and bad questions, are exempted from subjective judgments and only confined in the scope of this research. Nevertheless based on the selection of data attributes for characterizing good questions from the past literature, a selection of features that are used for our experimentation are shown in Table 1. A total of 21 attributes are used in this study.

Table 1. Features for characterizing high-quality questions

Category	Sub-category	Features
Meta data features	Asker's user profile	Reputation
		Days since joined
		Upvotes
		Downvotes
		Upvotes/Downvotes
		Questions asked
		Answers posted
		Answers posted /questions asked
	Question	Time
		Day
Content Features	Textual features	Tags
		Title length
		Question length
		Code snippet
		Wh word
	Content appraisal	Completeness
		Complexity
		Language error
		Presentation
		Politeness
		Subjectivity

2.2   Classification Algorithms

In this experiment, classification algorithms of machine learning are used to investigate the usefulness of the identified features to predict good and bad question in Stack Overflow. The performances of the popular classification algorithms are investigated: logistic regression, support vector machine (SVM), decision tree, naïve Bayes and k-Nearest Neighbors. These algorithms are implemented in the machine learning platform called Scikit-learn (http://scikit-learn.org). Scikit-learn is a free open source Python module integrating a wide range of state-of-the-art machine learning algorithms, with the emphasis on ease of use, performance, documentation completeness and API consistency [6]. The classification algorithms used are briefly explained below.

Logistic regression: Logistic regression applies the maximum likelihood estimation after transforming the features into a logistic regression coefficient. In this study, the logistic regression classification algorithm is applied with its L1 regularization parameter enabled in the feature selection process as it was found to be insensitive to the presence of irrelevant features. The logistic regression model is written as [7]:

$$p(y=1\,|\,x;\ \theta)=\frac{1}{1+\exp(-\theta^T x)} \tag{1}$$

where θ are the parameters of the model.

In the regularized logistic regression, obtain θ that solves the following optimization problem:

$$\arg\max_{\theta}\sum_{i=1}^{m}p(y^{(i)}\,|\,x^{(i)};\theta)-\alpha R(\theta) \tag{2}$$

where $R(\theta)$ is a regularization term that is used to penalize large weights/parameters. If $R(\theta)\equiv 0$, then this model is the standard, unregularized, logistic regression model with its parameters estimated using the maximum likelihood method. If

$R(\theta)\equiv\|\theta\|_1=\sum_{i=1}^{n}|\theta_i|$ then this is the L_1 regularized logistic regression.

Support Vector Machine (SVM): SVM is considered as one of the best classification algorithm for many real world tasks, mainly due to its robustness in the presence of noise in the dataset used for training, and also high reported accuracy for many cases.

Decision Tree: The classification and regression tree CART algorithm of the decision tree implemented in the Scikit-learn module are used. One of the benefits of using decision tree for classification is the ease of interpretability of the models and results. However, if there are irrelevant features in the training data, decision tree algorithm can create overly complex trees used for classification and causing over fitting (do not generalize the training data to unknown new data).

Naïve Bayes: One of the advantages of naïve Bayes is that it requires less training data to perform a classification compared to other algorithms. Its main disadvantage is that it cannot learn the interactions between the features, because a particular feature is assumed to be unrelated to other features, as mentioned earlier. In addition, naïve Bayes can suffer from oversensitivity to redundant or irrelevant features [8].

K-Nearest Neighbors (k-NN): In the testing phase, k is a user-defined constant, and a data point is classified by a majority vote of its neighbors, the class assigned is the class most common among its k nearest neighbors. The classification is highly depending on the number of nearest neighbor, k. Euclidean distance is a commonly used distance metric to find out the k nearest neighbors of a data point in the data space [9]. For this particular dataset, k=55 is found to give optimal classification results.

In the experiment, the stratified 10-fold cross-validation approach was used. Two evaluation metrics are adopted in validating the performance of the classification, the two evaluation metrics are: accuracy and area under the receiver operating characteristic (ROC) curve. In classification, accuracy is used to measure of the performance of binary classification test, the correctness of the algorithm in identifying or excluding a condition; the accuracy is the fraction of correctly classified results (both true positives and true negatives) of the overall results. The formula for accuracy is shown in equation (3), where TP is true positives: the number of positive classes that are classified as positives and TN is the true negative: the number of negatives classes that are classified as negatives. In the study, both positive class (questions with an accepted answer by the asker) and negative class (questions that are completely ignored) are equally important classes to be considered in the evaluation of the classification performance. Therefore, the evaluation metric: accuracy is used because it considers both true positives and true negatives in the measurement, unlike precision measure that only take the true positives into account.

$$accuracy = \frac{TP + TN}{TP + FP + FN + TN} \tag{3}$$

In addition to accuracy, the area under the ROC curve (AUC) is also used as an evaluation metric for a reliable comparative study. ROC is a curve that represents the performance of a classification model with the true positive rate (TPR) and false positive rate (FPR) when the discrimination threshold is varied. The TPR is the fraction of correct positive results to all the positive samples, whereas FPR is the fraction of incorrect positive results to all the negative samples. An ROC curve is a two-dimensional representation of the performance of a classification model. Similarly, AUC, the calculated area under the ROC curve, can be used to measure the performance of classification models. The AUC is statistically useful in the sense that the AUC is equivalent to the probability that the classification model will rank a positive class higher than a negative class that is randomly chosen.

2.3   Factor Analysis

Principal component analysis (PCA) is a popularly used method for extracting factor extraction as an exploratory type of factor analysis. It attempts to identify complex interrelationships among the attributes and the combinations of attributes that contribute to inducing a classified concept.

Factor weights are calculated in order to select the maximum possible variance, followed by further factoring continuing until there is no more meaningful variance remains. This approach helps summarize the variances in the dataset characterized by many attributes. Each attribute represents a different dimension. It could be difficult for a human to visualize a multi-dimensional hyperspace of attributes greater than three.

In essence the goal of PCA is to transform the original attributes into a new set of variables which reflect the variation in the data. These new variables that are formed corresponding to a linear combination of the initial attributes and are called principal components.

In this experiment, PCA is used to first reduce the dimensionality of the attributes of the data for constructing a classification model; and then the selected attributed are visualized graphically with minimal loss of information. Users can visually inspect the importance of each of these selected attributes pertaining to the predicated class. The correlation between the classification model attributes are calculated, and visualized as a correlation matrix via a correlogram in Figure 1.

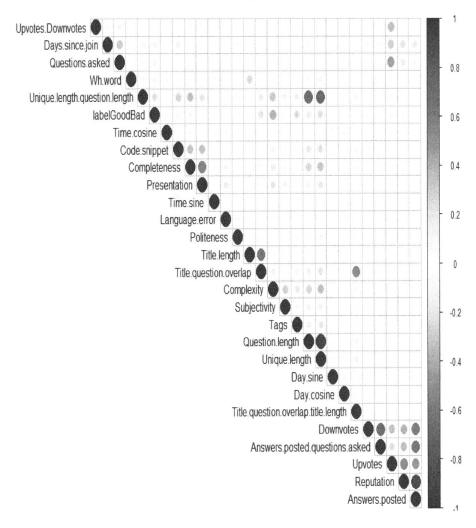

Fig. 1. The correlation matrix among the model attributes

The dataset is then processed by R with the built-in function of PCA. The generated eigenvalues are corresponding to the extent of the variation reflected by each principal component (PC). Eigenvalues should be strong for the first PC and their sizes shrink for the subsequent PCs. As shown in Figure 2 the first two eigenvalues constitute to almost 25% that indicates the first few PC account for sufficient variance.

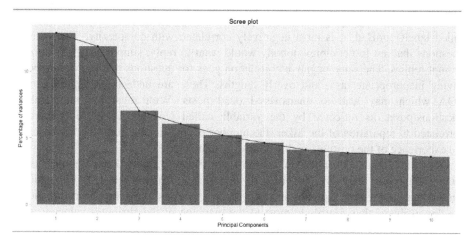

Fig. 2. Plot of eigenvalues vs Principal components

A R-based visualization package called FactoMineR (http://factominer.free.fr/) is used to chart up the cos2 on a Factor Map which is shown in Figure 3. The most

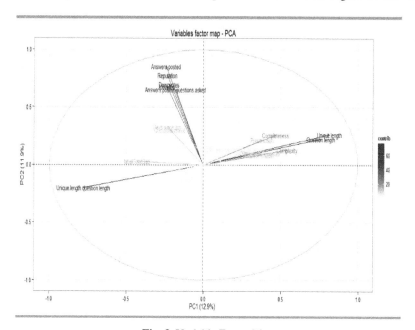

Fig. 3. Variable Factor Map

important variables in the determination of the principal components are highlighted in the hue of red, gradually down to blue.

From the correlation matrix, which is shown in Figure 1, there are two negatively correlated attributes, Unique questions and Question lengths. That shows questions that are unique but they are not long in sentences and vice-versa. Concise questions are mostly unique; long questions are duplicated or cross-posted in different places. When it comes to judging whether a question is a good question, by the variable called labelGoodBad, it is most negatively correlated with complexity. That means questions that are too complex nobody would want to reply; simple questions attract prompt replies. The same negative correlation goes for questions that are overlapped, having inappropriate tags, and overly lengths. These are undesirable features in a CQA which may lead to unanswered dead posts. With regards to favourable questions/posts, as reflected by the variable called Upvotes, they are positively correlated to reputation of the asker, the number of questions asked (that translate to the experience of the posters), and the days since joined.

From the Variable Factor Map, as shown in Figure 3, several most contributing factors stand-out; they are, in order of importance: Answers-posted, Reputation, Unique question length, Upvotes and Downvotes. In terms of machine learning, these are the attributes that contribute most to the mapping of the attributes values to the predicted classes. They contribute to the prediction power by their relative significances. The results can be interpreted that usually experienced posters who have a reputation, who post unique and concise questions would likely invite replies from the peer users. And their questions are likely to be voted, either up or down, implying certain popularity is attracted by these questions.

2.4   Classification Model Performance

The validation of the performance of classification is used to verify the usefulness and reliability of the attributes to predict good and bad questions in Stack Overflow. Table 2 shows the average accuracy and AUC from the stratified 10-fold cross-validation for all the classification algorithms used, containing both the average and AUC without feature selection (using the whole set of attributes) and with feature selection by PCA.

The value of the accuracy and AUC has a maximum value of 1 if the prediction of good and bad questions from the classification model is 100% correct, and a minimum value of zero if all the predictions are wrong. For a binary classification task in the study, the value of accuracy or AUC for a random guess is 0.5, which means that the classification model is useless if either the value accuracy or AUC falls under 0.5.

There are three important information that can be depicted from the results of the classification performance in Table 2. First, it is found that all the classification algorithms perform reasonably well in the prediction of good and bad questions. Without the feature selection, the overall accuracy is found to be ranging from 0.574 (naïve Bayes) to 0.735 (logistic regression). The AUC ranges from 0.759 (naïve Bayes) to 0.816 (logistic regression and SVM). This basically means that it is a feasible option to make use of the identified features to determine whether the questions are good or bad in Stack Overflow.

Secondly, the overall performance of the classification improves by replacing the features with a smaller set of features obtained from the feature selection step. With the inclusion of feature selection, the lowest accuracy actually improves to a 0.698 (k-NN) and the highest accuracy stays the same at 0.735, whereas the lowest AUC increases to 0.763 (k-NN). Among all the classification algorithms, the performance of naïve Bayes improves significantly with feature selection. This is because as mentioned earlier, naïve Bayes is sensitive to redundant or irrelevant features [8]. Therefore, this essentially means that the feature selection step successfully determines a subset of highly relevant and significant features, which serves a better representation of the dataset compared to using all the original features.

Thirdly, two classification algorithms, namely: logistic regression and SVM are found to have the best overall performance both in terms of accuracy and AUC, when compared to other algorithms. This is expected because these classification algorithms represent some of the best performing supervised learning methods in the current age [10].

Table 2. Average accuracy and AUC from 10-fold cross-validation.

Algorithms	Without feature selection		With feature selection	
	ACC	AUC	ACC	AUC
Logistic regression	0.735	0.816	0.735	0.813
SVM	0.734	0.816	0.735	0.813
Decision tree	0.728	0.780	0.732	0.784
Naïve Bayes	0.574	0.759	0.712	0.777
k-NN	0.694	0.763	0.698	0.763

3   Conclusions

Factor analysis was used to extract component of attributes and these components are used to build a classification model for classifying questions between the good (questions that contain at least an accepted answer by the askers) and bad questions (questions that are completely ignored by the community) in a CQA, with a case of topic Java in Stack Overflow.

The outcome of this study covers computational techniques for quantitatively finding the important features that are useful in the classification of good and bad questions. The features that are revealed from this study were extracted from the metadata of the askers' user profile and questions, as well as from the contents of the questions, which include textual features and content appraisal features.

From the analysis, it was found that when predicting the quality of a question, a user does not consider the previously asked similar or exactly same questions that are solved or unsolved. Information about the previously asked similar questions can be useful in classifying whether the newly asked question whether it is good or bad. In this way, the forum will be improved by providing guidelines on how to post questions that are likely to be answered. Likewise forum rules can be established for preemptively discouraging questions of poor styles to be posted to the forum. This improvement can be possible only when the insights between good and bad questions in CQA is known. Machine learning and feature selection methods offer such possibility.

Acknowledgement. The authors of this paper would like to thank Research and Development Administrative Office of the University of Macau, for the funding support of this project which is called "Building Sustainable Knowledge Networks through Online Communities" with the project code MYRG2015-00024-FST.

References

1. Li, B., Jin, T., Lyu, M.R., King, I., Mak, B.: Analyzing and predicting question quality in community question answering services. In: Proceedings of the 21st International Conference Companion on World Wide Web, pp. 775–782. ACM, April 2012
2. Chen, L., Zhang, D., Mark, L.: Understanding user intent in community question answering. In: Proceedings of the 21st International Conference Companion on World Wide Web, pp. 823–828, April 2012
3. Anderson, A., Huttenlocher, D., Kleinberg, J., Leskovec, J.: Discovering value from community activity on focused question answering sites: a case study of stack overflow. In: Proceedings of the 18th ACM SIGKDD International Conference on Knowledge Discovery and Data Mining, pp. 850–858 (2012)
4. Barua, A., Thomas, S.W., Hassan, A.E.: What are developers talking about? An analysis of topics and trends in stack overflow. Empirical Software Engineering, 1–36 (2012)
5. Mamykina, L., Manoim, B., Mittal, M., Hripcsak, G., Hartmann, B.: Design lessons from the fastest q&a site in the west. In: Proceedings of the SIGCHI Conference on Human Factors in Computing Systems, pp. 2857–2866, May 2011
6. Pedregosa, F., Varoquaux, G., Gramfort, A., Michel, V., Thirion, B., Grisel, O., Duchesnay, E.: Scikit-learn: Machine learning in Python. The Journal of Machine Learning Research 12, 2825–2830 (2011)
7. Ng, A.Y.: Feature selection, L 1 vs. L 2 regularization, and rotational invariance. In: Proceedings of the Twenty-First International Conference on Machine Learning. ACM, July 2004
8. Ratanamahatana, C.A., Gunopulos, D.: Scaling up the naive bayesian classifier: using decision trees for feature selection. In: Proc. Workshop Data Cleaning and Preprocessing (DCAP 2002), at IEEE Int'l Conf. Data Mining, ICDM 2002 (2002)
9. Weinberger, K., Blitzer, J., Saul, L.: Distance metric learning for large margin nearest neighbor classification. Advances in Neural Information Processing Systems 18, 1473 (2006)
10. Caruana, R., Niculescu-Mizil, A.: An empirical comparison of supervised learning algorithms. In: Proceedings of the 23rd International Conference on Machine Learning, pp. 161–168. ACM, June 2006

Data Projection Effects in Frequent Itemsets Mining

Mohammad Arsyad Mohd Yakop, Sofianita Mutalib,
and Shuzlina Abdul-Rahman

Faculty of Computer and Mathematical Sciences,
Universiti Teknologi MARA, 40450 Shah Alam, Malaysia
arsyadyakop@live.com,
{sofi,shuzlina}@tmsk.uitm.edu.my

Abstract. Nowadays, there are a number of algorithms that have been proposed in frequent itemsets mining (FIM). Data projection is one of the key features in FIM that affects the overall performance. The aim is to speed up the searching process by rearranging the items in a more compact form and to fit all the items in the data set in main memory efficiently without losing any information. The data refer to how the data set is stored in the main memory before the mining process begin. This paper explores the effects of data projection on frequent itemset mining from three different data projection types which are FP-Tree (tree-based), H-Struct (array-based) and FP-Graph (graph-based). The time construction and memory consumption are used to evaluate the parse and the dense of the data set. The result showed the construction of H-Struct is the fastest, but it consumes more time to mine frequent itemsets compared with FP-Tree and FP-Graph.

Keywords: array-based, data projection, graph-based, tree-based, frequent itemset mining.

1   Introduction

Frequent Itemsets Mining (FIM) involves searching and discovering process to find common items in the data set. The FIM is derived from a problem known as market basket analysis [1]. An item is frequent if the item's occurrence count in the data set is equal or greater than threshold called minimum support. The possible itemset combination in each data set is 2n-1, where n is the number of items in the data set. That means, the number of frequent itemsets in the data set are ridiculously large when the number of items in its increase. Therefore, frequent closed itemsets and frequent maximal itemset are the alternative ways to mine instead of normal frequent itemset. Frequent closed itemset is considered closed when there has no superset with the same support [2], [3] and frequent maximal itemset is determined when there is no superset that is frequent itemset.

Nowadays, the proposals of new algorithms are motivated by the data set that are expanded dramatically. Several algorithms have been proposed [4]–[11] in orders to have better efficiency either in aspect of memory space consumption or time

© Springer Science+Business Media Singapore 2015
M.W. Berry et al. (Eds.): SCDS 2015, CCIS 545, pp. 23–32, 2015.
DOI: 10.1007/978-981-287-936-3_3

consuming. The Apriori algorithm [1] is a classic frequent itemsets mining algorithm which is still popular till now. Candidate generation algorithm like Apriori and Carpenter requires a lot of memory consumption and slows down the searching performance since it enumerates all possible combinations of itemset [21].

In section 2, we describe the three types of data projection with an example for each type. In section 3, the experimental setup and the result are presented and in section 4, the discussion of the result was elaborated. Finally, we conclude with the summary of this in section 5.

2   Data Projection in FIM

Mining runtime and memory consumption are the two important factors in order to increase the performance of frequent itemset mining algorithm [11]. One aspect that researchers usually look into is to create a new data projection by rearranging the item in the data set so it will be mined faster as well as consuming low memory. Data projection is used to load the data set into the main memory in a compact form. Reviewing literatures [8], [12]–[15], there are three common types of data projections that were used to project the data set into the main memory which are Array-based, Tree-based, and Graph-based. Most proposed algorithms introduce the novel data projection that is more efficient in calculating the frequency and enumeration of the itemset.

2.1   Array-Based

LCMfreq [8] and HMiner [14] are the examples of algorithms that adopt the array-based structure. Basically, the array-based structure uses array data type to store the data set. With this structure, there is no complex construction to compress the data set into a compact form. LCMfreq and HMiner only use a standard array and not tree-based structure because they are not required to search the transaction in the data set. Moreover, array-based does not require additional initialization unlike tree-based and graph-based where it needs to be constructed first. This initialization sometimes needs more than one data set scanning.

H-Struct is proposed in Hmine [14] algorithm. The H-Struct uses a simple array structure to store data set in main memory. The H-Struct consists of two main parts which are header table and the list of arrays that hold the transaction information called frequent-item projection. The construction of an H-Struct started by finding frequent 1-itemsets in the data set. After all frequent 1-itemsets is found, the header table is created. Header table is a list of frequent 1-itemsets found in the data set and consists of three elements which are item-id, a support-count and hyper-link. This hyper-link in header table is used to point to the first item that occurs in the transaction that has the same item-id. For example, from Table 1 and illustrated in Fig. 1, the entry of item a in the header table H is pointing to the first item of transaction 200 in frequent-item projection. For transaction ID 200, the hyper-link of an item a is now pointed to the next occurrence of the item which is at the transaction ID 300.

Table 1. The transaction database TDB used as our running example [14].

Transaction ID	Items	Frequent-item projection
100	c, d, e, f, g, i	c, d, e, g
200	a, c, d, e, m	a, c, d, e
300	a, b, d, e, g, k	a, d, e, g
400	a, c, d, h	a, c, d

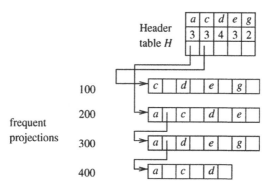

Fig. 1. H-Struct, the hyper-structure to store frequent-item projections [14].

2.2   Tree-Based

The Tree-based structure has been implemented in most proposed algorithms. The most famous and well-known is the FP-tree that has been introduced in FP-Growth algorithms [13]. FP-tree has shown the success in representing the data set to the compact form where it shares the common prefix among itemsets. This can be done by finding frequent 1-itemset in the data set and sorting it with the order of their frequency in descending order [16]. The list of frequent 1-itemset found is called the header table. Even though novel tree-based structure introduced nowadays [17], [18] are claiming more compact in representing the data set, most of them are the modification of the FP-Tree. Moreover, FP-Tree gets the compact representation by putting the items with higher frequency nearest to the root of the tree. However, the compression of the data set is working well in the dense data set, but not in sparse data set FP-tree [14]. This weakness of tree-based structure is because it will create more than one node for a single item and this will cause a higher memory consumption.

As an example of understanding how the construction of the FP-Tree, let Table 2 as the transaction data set DB and the minimum support threshold is set as 3. Before construction of the FP-tree starts, the data set was scanned to find frequent 1-itemsets and sorted by its frequency descending order. This list is named as header_table as shown in Table 3. Then, FP-tree is initialized and the root is marked as "null". The data set is scanned for a second time to construct the FP-tree and it begins to read all transactions in the data set and select frequent 1-itemsets from the each transaction.

Before frequent 1-itemsets is inserted it into the FP-tree, the items are sorted follows the header_table order.

Take the first transaction {*f, a, c, d, g, i, m, p*} as an example, the frequent 1-itemsets from this transaction is {f, a, c, m, p} and then it is sorted as per header_table order {*f, c, a, m, p*}. The first node inserted into the FP-tree is item *f* and its node_count is set to 1 as a child of the root, then the next item which is *c*, is stored as a child of the previous item which is item *f*. The process is repeated until all items in the first transaction is stored in the FP-tree. Going through the same process as the first transaction which has identified its frequent 1-itemsets, and with sorting transaction, the second transaction after the process will get {*f, c, a, b, m*}. Since the node of the first item in the second transaction has already created, then it only needs to increase the node_count with 1. However, for item *b*, there is no node *b* that exists as a child of node *c*. Then, the node b is created under node *c* and the node_count is set to 1 as illustrated in Fig. 2. This process is repeated until all transactions have been.

Table 2. Transaction Data set [13].

Transaction ID	Items	Ordered Frequent Items
100	*f, a, c, d, g, i, m, p*	*f, c, a, m, p*
200	*a, b, c, f, l, m, o*	*f, c, a, b, m*
300	*b, f, h, j, o*	*f, b*
400	*b, c, k, s, p*	*c, b, p*
500	*a, f, c, e, l, p, m, n*	*f, c, a, m, p*

Table 3. Header Table from DB

Item	f	c	a	b	m	p
Frequency	4	4	3	3	3	3

(a)   Insertion after the first transaction   (b) Insertion after second transaction

Fig. 2. Inserting items into FP-tree [13].

2.3   Graph-Based

The final method studied in this paper is based on its robustness in representing complex problem. The graph is defined as G = (V, E), where V is vertices (nodes) and E, is the edges that connect between vertices. Each vertex in the graph represents an item while the shows the relationship between the items. There are two types of graph theories that have been applied, which are Bipartite Graph and Directed Graph. The graph-based structure can compress the data set into a much smaller representation. The vertex created in the graph represents an item in a data set and the edge as the frequency of the two items connected. With that properties, vertices created in graph supposed to be less than tree-based structure. Even though the graph-based structure has shown its effectiveness in processing data sets with huge dimensional size, it still consumes more memory [19].

There are two sections involved in constructing FP-Graph. Firstly, the graph is constructed from the data set. Construction is performing by reading each item in each transaction in the data set. It begins by creating the node for the first item in the first transaction and set it to 1 as the frequency value. Then the iteration is moved to the second item and create the node. Between the first and second node, the edge is created to connect these two nodes. This process is repeated until the last item in the transaction is enumerated. After finishing the first transaction, construction graph now moved to the second transaction and again the first item is selected. If the node of the first item has already created, the frequency of the node is increased by 1, but if it's still does not exist in the graph then the node is created. The process is repeated until all items in all transactions are enumerated.

The FP-Graph then needs to be pruned to remove all items with a frequency below the threshold. While removing the infrequent nodes, if there is an edge that exist between infrequent and frequent node, the edge that source from the infrequent node needs to be removed. Then, the edge that pointing to the infrequent node are modified to point to the removed edge destination node. Then the modified edge needs to check if there is another edge that equals with this edge. If there is a same edge, one edge need to be deleted and another one needs to increase its edge frequency. An example of the pruning process is as illustrated in **Error! Reference source not found.**.

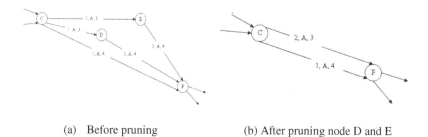

(a)   Before pruning   (b) After pruning node D and E

Fig. 3. Pruning in a FP-Graph [15]

3   Experimental Setup

This experiment examined three data projection types: array-based (H-Struct), tree-based (FP-Tree) and graph-based (FP-Graph). The experiment is done using Intel i5 CPU at clock rate 3.40GHz and 4GB (3.80GB usable) memory. In order to evaluate these three types of data projection, HMiner, FP-Growth and FP-Growth-Graph algorithms have been selected and it has been implemented in Java. Implementation of FP-Growth and HMiner algorithms are taken from SPMF library, GPL v3 license [20]. For FP-Growth-Graph, the implementation uses the SPMF library.

The data set used in this experiment is an Accident and Retail were both acts as an example of a dense and sparse type of data set respectively. These data sets are widely used in testing the frequent itemsets mining algorithms [8]. The characteristics of the data set as shown in Table 4.

Table 4. Data set Characteristics

Data set	Transactions	Items	Avg Items/ Transaction	Category
Accident	340,183	468	33	Dense
Retail	88,162	16,469	10	Sparse

4   Result

4.1   Construction Time

The execution time taken into consideration in this study is the time taken for the algorithm to construct the data projection in the main memory. The effectiveness of the data projection process before the mining process starts is investigated. For dense data set in Fig.4 (a), both FP-tree and FP-graph has shown that the higher support threshold, the less time taken to construct tree/graph projection. This is because when the threshold value is higher, less number of frequent 1-itemsets is selected which made the construction become more faster. Since the FP-graph has only created 1 node for an item, it gives an advantage to FP-graph to performing the construction faster compared to the FP-tree that needs to find existing child node for the current item before it can process either to create new node or update the existing node. As a result, H-Struct is taking longer time as it needs to copy all the items in each transaction that above the threshold and enumerates the items one by one since it have higher average items per transaction. Nevertheless, the three data projection types shown in gives similar results when it is tested on sparse data set as shown in Fig.4 (b). There are only slight differences between these three data projections. Thus, it can be concluded that all the three methods are performing equally when the number of frequent 1-itemset selected is small. The result of the data projection experiment is shown in Fig. 4 and Fig. 5.

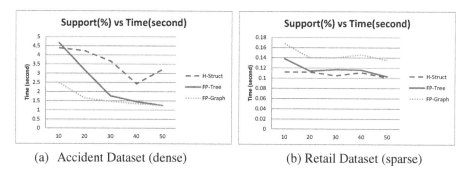

(a) Accident Dataset (dense)   (b) Retail Dataset (sparse)

Fig. 4. The data projection's construction time.

4.2   Execution Time

Figure 5 shows the execution time for dense and parse data set for FP-Growth, FP-Growth-Graph and HMine algorithms. HMine clearly take more time in mining frequent itemsets on array-based data projection compared with the other data sets. HMine run slower since the algorithm needs to process multiple times for the same transaction. However, for the dense data set, FP-Growth-Graph performs much better than FP-Growth but in sparse dataset, it is otherwise. The ability of FP-Tree in compressing the dataset into compact form gives the advantage for FP-Growth in mining denser data set compared to HMine but not FP-Growth-Graph. Our investigation has shown that to mine frequent itemset from FP-Graph, it needs to start from all nodes in the graph. Starting from initial node, searching process used Depth First Search manner to find all its prefixes. In comparison with FP-Tree, it uses simpler way to find the prefix of an item by a traverse to its parent toward the root. This is why the FP-Growth takes less execution time than FP-Growth-Graph in the sparse data set since the FP-Graph has many possible paths to traverse. This results have demonstrated that data arrangement is one of the factors that influence the efficiency of the algorithms.

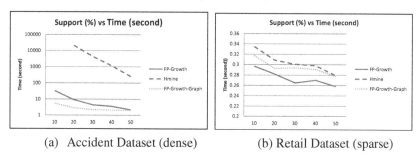

(a) Accident Dataset (dense)   (b) Retail Dataset (sparse)

Fig. 5. The algorithms excution time.

4.3   Memory Usage

Another factor evaluated in this experiment is the data projection memory usage. Hence, the memory usage is recorded and shown in Fig. 6. The result obviously shown that FP-Graph required less memory compared to the other two data projections. In dense data set, there is obviously graph changes in FP-Graph when the support threshold increase from 20% to 30%, where it shows the differences frequent 1-itemsets found in the data set. For sparse data set FP-Graph outperform FP-Tree and H-Struct.

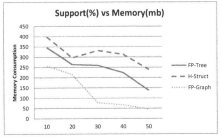
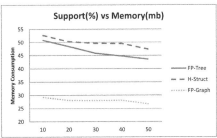

(a) Accident Dataset (dense)   (b) Retail Dataset (sparse)

Fig. 6. The data projection's memory consumption.

In term of the memory usage for both data sets, as shown in Fig. 7, HMine is taking a lead after record the high memory consumption compares to FP-Growth and FP-Growth-Graph. Since it applied the array-based data projection, the data were constructed in the memory is based on how many it items appears in the data set. When H-Struct data projection that applied HMine consumes more memory, the mining time also will be affected since it giving a burden to the hardware to process especially for the dense data set. It is shown FP-Growth-Graph that applied graph-based data projection uses less memory for dense and sparse data set. This is because of the FP-Graph does not duplicate the node for an item or for simple understanding, one node is dedicated for one item. FP-Growth have similar memory consumption with

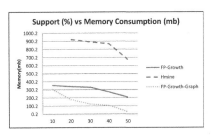
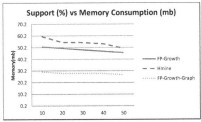

(a)   Accident Dataset (dense)   (b) Retail Dataset (sparse)

Fig. 7. The algorithms memory consumption.

FP-Growth-Graph as it applied the common prefix sharing. This prefix sharing is to compact the representation of the data projection. For sparse data set, the memory consumption is similar to HMine because the FP-Tree cannot compress the sparse dataset efficiently. This is because there is less common prefix to share due to the contained items in each transaction and lead the duplication of the node.

5   Conclusion

This paper evaluated three types of data projection: array-based, tree-based and graph-based that were applied on three different algorithms. H-Struc in HMine algorithm works well with sparse data set, but it took longer construction time in the dense data set. The advantage of H-Struct array-based projection is that it uses a simple array and store the data set in main memory in its original form. The straight forward implementation of H-Struct in HMine makes is easy to understand and suitable to be used on the sparse data set. However, the use of HMine in dense data depends on the capability of the hardware. The FP-Tree works well in both types of data set. The compression and arrangement of the data set showed that the tree-based projection can be used in both dense and sparse data set. In this experiment, FP-Graph has shown an outstanding performance for all the tested data. Similar to FP-Tree, FP-Graph reduces the node creation to minimize the memory usage and as well as construction time.

References

1. Agrawal, R., Srikant, R.: Fast algorithms for mining association rules. In: Proc. 20th Int. Conf. Very Large Data Bases, VLDB, vol. 1215, pp. 487–499 (1994)
2. Zaki, M.J.: 1 Mining Closed & Maximal Frequent Itemsets. NSF CAREER Award IIS-0092978, DOE Early Career Award DE-FG02-02ER25538, NSF grant EIA-0103708
3. Hadzic, F., Tan, H., Dillon, T.S.: Mining maximal and closed frequent subtrees. In: Hadzic, F., Tan, H., Dillon, T.S. (eds.) Mining of Data with Complex Structures. SCI, vol. 333, pp. 191–199. Springer, Heidelberg (2011)
4. Patel, S., Kotecha, K.: A Prime Number based framework for Frequent Pattern Mining (2014)
5. Chai, D.J., Jin, L., Hwang, B., Ryu, K.H.: Frequent pattern mining using bipartite graph. In: Database and Expert Systems Applications, pp. 182–186 (2007)
6. Shenoy, P.D., Srinivasa, K., Thomas, A.O.: Compress and Mine: An Efficient Graph Based Algorithm to Generate Frequent Itemsets (2004)
7. Lucchese, C., Orlando, S., Perego, R.: DCI closed: A fast and memory efficient algorithm to mine frequent closed Itemsets. In: FIMI (2004)
8. Uno, T., Kiyomi, M., Arimura, H.: LCM ver. 2: Efficient mining algorithms for frequent/closed/maximal itemsets. In: FIMI, vol. 126 (2004)
9. Pan, F., Cong, G., Tung, A.K., Yang, J., Zaki, M.J.: Carpenter: Finding closed patterns in long biological datasets. In: Proceedings of the Ninth ACM SIGKDD International Conference on Knowledge Discovery and Data Mining, pp. 637–642 (2003)

10. Zaki, M.J., Hsiao, C.: Charm: an efficient algorithm for closed association rule mining (1999)
11. Pyun, G., Yun, U., Ryu, K.H.: Efficient frequent pattern mining based on linear prefix tree. Knowledge-Based Systems 55, 125–139 (2014)
12. Moonesinghe, H.: Graph based methods for pattern mining. ProQuest (2007)
13. Han, J., Pei, J., Yin, Y.: Mining frequent patterns without candidate generation. ACM SIGMOD Record 29(2), 1–12 (2000)
14. Pei, J., Han, J., Lut, H., Nishio, S., Tang, S., Yang, D.: H-Mine: Fast and space-preserving frequent pattern mining in large databases. IIE Transactions 39(6), 593–605 (2007)
15. Tiwari, V., Gupta, S., Tiwari, R.: Association rule mining: A graph based approach for mining frequent itemsets. In: 2010 International Conference on Networking and Information Technology (ICNIT), pp. 309–313 (2010)
16. Roy, S.: Frequent Pattern Itemset Mining For Big Data Using FP-Tree Approach (2014)
17. Cheung, W., Zaiane, O.R.: Incremental mining of frequent patterns without candidate generation or support constraint. In: Proceedings ot Seventh International Database Engineering and Applications Symposium, pp. 111–116 (2003)
18. Schlegel, B., Gemulla, R., Lehner, W.: Memory-efficient frequent-itemset mining. In: Proceedings of the 14th International Conference on Extending Database Technology, pp. 461–472 (2011)
19. Yen, S.J., Chen, A.L.P.: A graph-based approach for discovering various types of association rules. IEEE Transactions Knowledge and Data Engineering 13(5), 839–845 (2001)
20. Fournier-Viger, P., Gomariz, A., Gueniche, T., Soltani, A., Wu, C.W., Tseng, V.S.: SPMF: a Java open-source pattern mining library. The Journal of Machine Learning Research 15(1), 3389–3393 (2014)
21. Mutalib S., Abdul-Rahman S., Mohamed, A.: Mining frequent patterns for genetic variants associated to diabetes. In: Database and Expert Systems Applications (DEXA), pp. 28–32, September 1–5, 2014

Data Quality Issues in Data Migration

Nurhidayah Muhamad Zahari[1], Wan Ya Wan Hussin[2],
Mohd Yunus Mohd Yussof[2], and Fauzi Mohd Saman[2]

[1] Faculty of Computer and Mathematical Sciences
Universiti Teknologi MARA
[2] Shah Alam, Selangor, Malaysia
sunputri89@gmail.com
{wanya,mymy,fauzi}@tmsk.uitm.edu.my

Abstract. The main criterion of a successful data migration project is the data quality. Quality of data can be compromised depending upon how the data are received, integrated, maintained, processed and loaded. The data migration project requires the data to be extracted from multiple sources before being cleansed and transformed. Once the data are cleansed and transformed, the data will be loaded into a new system. Therefore, data cleansing is the most important activity in a data migration project. Data cleansing is the process of detecting and removing errors, inconsistencies and redundancies in order to improve the quality of data

Keywords: data quality, data migration, extract, transform and load (ETL), data cleansing.

1   Introduction

Data migration is a process of transferring data from one system to other systems [1]. Whenever legacy data is involved, the data migration process becomes more complex. Legacy data is a collection of information that has been recorded over time [2], which contains business knowledge such as product data, customer details and so on. Since the legacy data is valuable to the organizations, any data migration must be complete, accurate, consistent, conformity and integrity manners.

Due to new requirements and specifications, new systems may use data structures, which are different from the legacy systems. Therefore, one-to-one mapping between new and existing systems may not be possible.

In any software project, data migration should emphasize on [3]:

- Ensuring that data is being delivered to target fits with the purpose, not only with respect to the technical requirements but also with respect to the behavior of the target system.
- Ensuring all data being extracted is relevant to the requirements and delivered without any breakages. Breakages mean data should not be lost, wrongly transformed or duplicated.
- Justifying reasons to why the source data has not been delivered to the target.

© Springer Science+Business Media Singapore 2015
M.W. Berry et al. (Eds.): SCDS 2015, CCIS 545, pp. 33–42, 2015.
DOI: 10.1007/978-981-287-936-3_4

In 2012, the Accountant General Department of Malaysia has adopted the International Public Sector Accounting Standards (IPSAS) as its new standard accounting practice. As a result, the new accounting and financial management system known as 1Government Financial and Management Accounting System (1GFMAS) was developed. The main objective of the 1GFMAS is to support both cash and accrual practice of accounting in a centralized environment [5].

Data migration is one of the 1GFMAS project scope. The main objective of data migration is to transfer all the legacy data from 37 sites of decentralized Government Financial Management and Accounting System (GFMAS) to 1GFMAS. To ensure business continuity, the legacy data, such as master data, balances and open transaction are transferred in complete, accurate, consistent, conformance and integrity manners specified.

2   Background of the Study

In this section, we will discuss about data migration method.

2.1   Data Migration Method

There are a number of methods used to perform data migration. The most commonly used methods are:

- Extract, Transform and Load (ETL) – Data from various sources is extracted and transformed to the new format according to the requirements before loading it to a new target system.
- Enterprise Application Integration (EAI) – This technology is the best to use if the amount of the data is small. The migrating process is fast between logic layers of application.
- Enterprise Data Integration (EDI) – Enterprise data integration is the combination of technical and business processes used to combine data from different sources into meaningful and valuable information. A complete data integration solution encompasses discovery, cleansing, monitoring, transforming and delivery of data from a variety of sources.
- Object Relational Mapping (ORM) - Object relational mapping is an incompatible technique for converting incompatible data. Object relational mapping makes it possible to address access and manipulate objects without having to consider how those objects relate to their data sources. ORM lets programmers maintain a consistent view of objects over time, even as the sources, the sinks and the applications have changed.

Of the four methods, ETL is the most popular method that use in data migration [4]. Based on the survey conducted by The Data Warehousing Institute (TDWI) [4], 41% of respondents prefer to use ETL than the other methods. This is due to the fact that its ability to handle the extreme requirements of data migration, including big data, multi-pass data transformations, deep data profiling, interoperability with data quality tools and many-to-many data integration capabilities.

2.2   Extract, Transform and Load (ETL)

There are three main processes in ETL, which are data extraction, data transformation and data loading. In the data transformation process, it has three sub-processes, which are data profiling, data cleansing and data mapping [6] [7] [8] [9] [10]. Each process will produce output files. The output files will only be loaded after the data mapping process has completed. The quality of data and the effectiveness of the data migration process are directly dependent on the efficiency of the ETL processes.

ETL consists of five main steps so that data migration process can be completed. These steps are further details as the following:

- Data Extraction – Data extraction is the first stage of the data migration process. It is a process of retrieving data from the legacy system. These data can be retrieved from multiple sources and formats, for example databases, flat files, spreadsheets, etc. The main objective of data extraction is to retrieve all required and relevant data from the source system within a certain time. The output from the data extraction process will be used in the data cleansing process.
- Data Profiling – Data profiling is the process to uncover the type and condition of the data available. The profiles will not only give detailed information about the data (usually referred to as files, columns or fields) but also the quality of the content and the value and structure of the data. The extracted data must go through the normalization process in order to avoid redundancy and duplication. The objective of data profiling is to equip project with a clear understanding of the data that will help ensure all situations are addressed.
- Data Cleansing - Data cleansing is also known as data scrubbing or data cleaning, which is part of the data transformation process. The main objective of this process is to remove errors and inconsistencies from the extracted data to improve the quality of the data before migrating them to the target system. In the ETL process, data cleansing is the most important and critical part.
- Data Mapping - After the data cleansing process, the cleansed raw data will be mapped and transformed into the final upload template. This process is called a data mapping or data transformation. The output data obtained from the data cleansing process is used as an input for the data mapping process. The transformation of data involves a series of rules based on business requirements.
- Data Loading – The last step of the ETL process is the data loading process where output from the data transformation process will be loaded to the target system.

2.3   Data Quality Issues

Data quality is the main factor to identify either data migration project is successful or not. Poor data quality was the main contributor (53.4%) to project overruns [11]. Poor quality of master data will unable to support business rule or process. A high level of risk is involved in data migration projects due to large or big data. The major challenge of the migration project is to ensure that the metadata is retained in its original form. If data with poor quality, inaccurate or inadequate is migrated, it can lead to inappropriate assumptions, misleading results and incorrect decision-making.

Ensuring high levels of data quality is the most expensive and time-consuming task in a data migration project. There are six major data quality problems [12] [13] [14] identified during the data migration process: data type mismatch, specification mismatch, missing values, typing errors, redundant data and incorrect data values.

2.4   Data Quality Framework

To have a successful data migration project, high quality data are required. The high quality data are defined as the data fulfilled six dimensions of data quality - completeness, consistency, validity, conformity, accuracy and integrity [18] [19] [20] [21] [22]. These goals should be the key driving factor in choosing the data quality framework.

Data quality framework is a process for improving data quality [18] [19] [20] [21] [22]. It is used to ensure the quality of the data at every stage of the data life cycle – from data capturing, data transformation and data usage. Such a pervasive framework will be easier to implement in new organizations that are defining and building their IT architecture from scratch. However, it is becoming complex and difficult when the involvement of multiple operational systems across multiple organizations.

The basic components of a data quality framework are assessment, cleansing and enhancement and consolidating (Fig 1).

Fig. 1. A Simplified Data Quality Framework

- Assessment - is a process that analyses, inspect, and measure the degree of quality of the data before determining the data cleansing and enhancement strategies.
- Cleansing and Enhancement - based on the assessment finding, the source data will be parsed, standardized and corrected. Wherever necessary, the cleansed data will be enriched to add additional data into it.
- Consolidation - the duplicate data are removed so that the data matching could be done to determine the records are referring to the same entity. Once everything is finished, the data migration can be started.

3   Results

In this section, discussions about data quality issues or problems in data migration based on case study are laid out.

3.1   Data Type Mismatch

The mismatch data types caused problems during extraction and loading processes. The definition of data types in computer programming is a classification of the types of the data for example, integer, string, character and so on.

Table 1. A sample data source that has a data type mismatch problem

	Source_1	Source_2
Technical Name	PERNR (PK)	PERNR (PK)
Bussiness Name	Personal Number	Personal Number
Data type	Integer	Varchar
Sample Value	2012170243	A2012170243

In Table 1 above, there are two data sources, which are Source_1 and Source_2. In Source_1, the Personal Number attributes are declared as an integer and in Source_2, they are declared as a string of type VARCHAR. The Personal Number of both data sources was chosen as the primary key respectively. The fusion of these records is hard to manage because while it is possible to insert integer values into a string field, integer field cannot accept an alphanumeric value.

3.2   Specification Mismatch

This is a problem of multiple abbreviations, for example, with three different data sources, the GENDER specification attributes are having the same data type, but were expressed differently as shown in the Table 2.

Table 2. A sample data source that has the specification mismatch problem

	Source_1	Source_2	Source_3
Technical Name	GENDER	GENDER	GENDER
Bussiness Name	Gender	Gender	Gender
Data type	Varchar	Varchar	Varchar
Sample Value	Male	M	L

In these data sources, the data type GENDER is declared as VARCHAR. However, in Source_1, the record for GENDER is specified as a 'Male', in Source_2 the record for GENDER is specified as 'M' and in Source_3 the record for GENDER is speci-fied as an 'L'. Different specifications of data in data sources make the extraction be-come more complex and the conversion to a standard format is needed.

3.3   Missing Values

Occasionally, some values of a field either in the source or the target are missing. For these missing data, developers will have to assign a default value. However, the as-signed default values sometimes may not be appropriate because of the problem of defining what the data is actually represents.

A null value of a record may have dependencies to other tables and this may further worsen the data quality problems. By not specifying the null values in the data sources, however, may result in generating incomplete or inconsistent data.

Table 3. A sample data source with missing value problem

SoureHR1

PERNR	GSBER	KOSTL	GEBER	FKBER
00082784	1106	0341201000	B27	B27050301
00111468	1106	0341104000	B27	010102
00525485	1106	0341267000	B27	040501
00558890	1106	0341207000	B27	B27050801
00727990	1106	0341101000	B27	B27010407
00753806	1106	0341216000	B27	030301
00764407	1106	0341202000	B27	050201
00764837	1106	0341202000	B27	050201
00764902	1106	0341202000	B27	050201
00791947	1106	0341231000	B27	030101

SoureHR2

PERNR	SNAME	SADDRESS	SDOB
00082784	NOR MAZLIAZRI	PUCHONG	22-12-1979
00111468	MOHD MUSTAQIM	NILAI	30-10-1986
00525485	HAZWAN HAKIM	SERI KEMBANGAN	04/06/1989
00558890	FAIZ SARONI	KUALA LUMPUR	31-07-1984
00727990	NUR FARAH HIDAYU	PUTRAJAYA	01/01/1990
00764407	MOHD FIRDAUS	KAJANG	14-03-1980
00764407	MAT AFIF ABDULLAH	CHERAS	02/03/1981
00764837	SITI RASHIDAH OMAR	AMPANG	00-00-0000
	TAN KAI XUAN	GOMBAK	30-02-1983
00791947	TENGKU NURUL HUDA	SUBANG	14-03-1984

In Table 3, for example, there are two data sources; SourceHR1 and SourceHR2 with the values of some attributes from both sources may be needed to migrate together into one output table. The Personal Number or PERNR is the key attribute to join these two data sources. As a result, some values of the key attribute were missing in SourceHR2 which produced an output table that contain 'null' values. This violates the principles of primary keys.

3.4   Typing Errors

This type of error is most common. Table 4 shows a sample of type errors that commonly occur in data source. Attribute GENDER is declared as Varchar in both data sources, Source_1 and Source_2. If the attribute value 'Male' is misspelled as 'amle' in Source_2, then during the migration, it will generate a complete but incorrect data.

Table 4. A sample data source that has a typing error

	Source_1	Source_2
Technical Name	GENDER	GENDER
Bussiness Name	Gender	Gender
Data type	Varchar	Varchar
Sample Value	Male	amle

3.5   Redundant Data

The presence of duplicate data from multiple sources may also cause data quality issues. Data duplication may cause inconsistent data retrieval, thus produce differing output.

Table 3 shows duplicate entries in the data source SourceHR2 for PERNR = '00764407'. When these two data sources are joined with PERNR is treated as join attribute, the result join conflict where it is not sure to choose which record from table SourceHR2. At the end, only one record for each unique PERNR is retained.

3.6   Incorrect Data Values

In Table 4 where the date of birth (SDOB column) attributes is having future date which is incorrect data. In some cases, the migration team needs to get back to the user, since only the user knows the details. The data migration can only start once the data is repaired and updated.

4   Findings and Discussion

Data migration has development cycle. In the case study, a pre-migration process known as Mock Run was performed. The Mock Run was executed three times - Mock Run1, Mock Run 2 and Mock Run 3. The last mock run was performed to ensure the remaining issues are resolved.

4.1   Enhancement of Data Cleansing Process

The most important part of the data migration project is data cleansing, which is to ensure only high quality data is transferred and used in the new system. This enhancement, which requires involvement of users, is based on processes below [15]:

- Develop job based on cleansing rules
- Correct the errors based on cleansing rules for each instance
- Produce validation report or rejection file and display to the user, so that they can make a correction manually.
- After data been corrected by the user, they will pass to migration team to check the rules again.
- This process will be looped until no more validation report is produced.
- Finally, the cleansed will be storage, temporary staging before proceeding to data mapping

In the case study, an algorithm was developed to perform cleansing rules for each scenario using IBM's INFOSPHERE, DATASTAGE and QUALITYSTAGE.

4.2   Data Type Mismatch

Data type conversion was performed before the data were mapped to target output using a tool which provides a type conversion function called *DecimalToString()*. This function returns any numerical value as a string.

However, the type conversion rules were performed only on Source1 because it contains problematic data. After the cleansing process was done, the Personal Number column is defined the data type as a *Varchar*.

4.3   Specification Mismatch

Multiple data specifications in data sources need to be converted to a standard format. Based on example in Table 4, Source1, Source2 and Source3 were using different specifications for the same attribute. A standardized specification is determined after discussing with the users.

If multiple data values are involved, *'If... Else'* statement is used. An example, in the output target, users want all the values for the gender attribute is *'L'*, which is for *'Male/M'* and *'P'* for *'Female/F'*. At the end, we will have standard values for the gender attribute, which is *'L'* for male, and *'P'* for female.

4.4   Missing Values

User involvement in solving the missing values is crucial since only they know what it is. As part of the migration process, validation report will be produced during data cleansing. This report will help user to identify which column is having data issues.

The first step that needs to be done as part of the cleansing process is creating 'rules' to check the missing values. For example, Personal Number (PERNR column) attribute should not contain blank or null values. It is mandatory to have a valid value where PERNR column is the primary key. Second step is to pass the rejected list to users. The users will have to fix their source table before data migration.

4.5   Typing Error

According to Gümbel [16], 76% of the data quality issues are caused by the users. Typing error is one of the most common mistakes made by user while entering the data. The solution is quite similar with specification mismatch. Since the possibility of having multiple types of data values is high, the *'If... Else'* statement is used.

Not all data quality issues can be resolved by data cleansing. The data migration team needs to get the users' involvement. In certain cases, the users will have to update the data. It is good to have user involvement in the data migration process. The users can cleanse and validate the data at the same time. Consequently, the quality of operational data sources can be improved [17].

4.6   Redundant Data

In the case study, many of data sources are having data redundancy issue. The primary key should not have multiple values. If this issue arises, then one of the records was retained and others were removed.

In Table 5, for example, there are duplicate entries in the SourceHR2 file for PERNR = '00764407'. Cleansing rule was created to solve this issue by removing duplicate entries.

Table 5. A sample output data after cleansing process

Unique SourceHR2			
PERNR	SNAME	SADDRESS	SDOB
00082784	NOR MAZLIAZRI	PUCHONG	22-12-1979
00111468	MOHD MUSTAQIM	NILAI	30-10-1986
00525485	HAZWAN HAKIM	SERI KEMBANGAN	04/06/1989
00558890	FAIZ SARONI	KUALA LUMPUR	31-07-1984
00727990	NUR FARAH HIDAYU	PUTRAJAYA	01/01/1990
00764407	MOHD FIRDAUS	KAJANG	14-03-1980
00764837	SITI RASHIDAH OMAR	AMPANG	00-00-0000
00764902	TAN KAI XUAN	GOMBAK	30-02-1983
00791947	TENGKU NURUL HUDA	SUBANG	14-03-1984

4.7   Incorrect Data Values

In Table 6, when the date of birth (SDOB column) attributes was having incorrect date, the users must correct them before data can be migrated successfully.

Table 6. A sample output after cleansing process

Valid Date in SourceHR2				
PERNR	SNAME	SADDRESS	SDOB	IND_DATE
00082784	NOR MAZLIAZRI	PUCHONG	22-12-1979	√
00111468	MOHD MUSTAQIM	NILAI	30-10-1986	√
00525485	HAZWAN HAKIM	SERI KEMBANGAN	04/06/1989	√
00558890	FAIZ SARONI	KUALA LUMPUR	31-07-1984	√
00727990	NUR FARAH HIDAYU	PUTRAJAYA	01/01/1990	√
00764407	MOHD FIRDAUS	KAJANG	14-03-1980	√
Invalid Date in SourceHR2				
PERNR	SNAME	SADDRESS	SDOB	IND_DATE
00764837	SITI RASHIDAH OMAR	AMPANG	00-00-0000	×
00764902	TAN KAI XUAN	GOMBAK	30-02-1983	×

We need to perform validation check during the data cleansing process. In our case study, we will check the validity of the record by using Invalid () function. A rejected table will be created if the record is having invalid date values. Then, the table will be given to the user to correct the date.

5   Conclusion

The data cleansing is a crucial activity in any data migration projects since the main criterion of a successful data migration. To improve the quality of data, data cleansing is performed to eliminate errors, inconsistencies and redundancies. By using a high quality data, the organization should be able to use the new system at fullest and become more effective and efficient organization than before.

References

1. Manjunath, T.N., Hegadi, R.S.: Data Quality Assessment Model for Data Migration Business Enterprise. International Journal of Engineering and Technology (IJET) 5(1), February-Marche 2013
2. Baylaan Technologies Inc., Is Data Migration Holding Your Company Back, Ontorio, Canada. http://www.baylaan.com/white_papers.shtml
3. X88 Software Limited, Data Migration Method (revision November 3, 2013)
4. Russom, P.: Best Practices in Data Migration. The Data Warehousing Institute, April 2006
5. Accountant General Department Malaysia, Transition Towards Accrual Accounting (2012)
6. Anand, N.: Application of ETL Tools in Business Intelligence. International Journal of Scientific and Research Publications 2(11), November 2012
7. Kakish, K., Kraft, T.A.: ETL Evolution for Real-Time Data Warehousing (2012)

8. El-Sappagh, S.H.A.: Journal of King Saud Uni., Optimizing ETL Processes in Data Warehouse (2011)
9. Patil, R.Y., Kulkarni, R.V.: A Review of Data Cleaning Algorithms for Data Warehouse Systems 3(5) (2012)
10. Rahm, E., Do, H.H.: Data Cleaning: Problems and Current Approaches. IEEE Computer Society Technical Committee on Data Engineering (2000).
    http://dbs.uni-leipzig.de/file/TBDE2000.pdf
11. Bloor Research, Data Migration White Paper, London, UK, May 2011.
    http://www.bloorresearch.com/research/white-paper/
    data-migration-white-paper/
12. Singh, J., Singh, K.: A Descriptive Classification of Causes of Data Quality Problems in Data Warehousing. International Journal of Computer Science 7(3), No 2, May 2010
13. Chillar, R.S., Kochar, B.: Research Problem in Data Warehouse. INDIACom. (2008)
    http://www.bvicam.ac.in/news/INDIACom%202008%20Proceedings/p
    dfs/papers/98.pdf
14. Singh, J., Singh, K.: Statistically Analyzing the Impact of Automated ETL Testing on the Data Quality of a Data Warehouse. International Journal of Computer and Electrical Engineering 1(4), October 2009
15. Hamad, M.M., Jihad, A.A.: An Enhanced Technique to Clean Data in the Data Warehouse (2011)
16. IBM, Data Quality – the Base for your Enterprise Applications, August 2007.
    http://www-01.ibm.com/software/
    sw-library/_US/detail/P284425Y67497T03.html
17. Bradji, L., Boufaida, M.: Open User Involvement in Data Cleaning for Data Warehouse Quality. International Journal of Digital Information and Wireless Communications (IJDIWC) (2011)
18. North Lincolcshire Council, Data Quality Frameworj Version 1.3 (2015).
    http://www.northlincs.gov.uk/_resources/assets/attachment/
    full/0/21694.pdf
19. Jerome, C., Pascal, R.: An Enterprise Architecture and Data quality framework (2013). Springer Link http://link.springer.com/chapter/10.1007%2F978-3-642-37317-6_7
20. GS1, Data Quality - An Introduction to the role and value o data quality in organisations (2010). http://www.gs1pa.org/pdf/importanciadatos.pdf
21. Shirlee-ann, K., Janice, B.: Developing a Framework for Assessing Information Quality on the World Wide Web. Informing Sciene Journal 8 (2005).
    http://inform.nu/Articles/Vol8/v8p159-172Knig.pdf
22. Karolyn, K.: The Development of a Data Quality Framework and Strategy for the New Zealand Ministry of Health.
    http://mitiq.mit.edu/Documents/IQ_Projects/Nov%202003/
    HINZ%20DQ%20Strategy%20paper.pdf

Reviewing Classification Approaches in Sentiment Analysis

Nor Nadiah Yusof, Azlinah Mohamed, and Shuzlina Abdul-Rahman

Faculty of Computer and Mathematical Sciences, University Teknologi MARA,
40450 Shah Alam, Selangor, Malaysia
nornadiah.yusof@gmail.com,
{azlinah,shuzlina}@tmsk.uitm.edu.my

Abstract. The advancement of web technologies has changed the way people share and express their opinions. People enthusiastically shared their thoughts and opinions via online media such as forums, blogs and social networks. The overwhelmed of online opinionated data have gained much attention by researchers especially in the field of text mining and natural language processing (NLP) to study in depth about sentiment analysis. There are several methods in classifying sentiment, including lexicon-based approach and machine learning approach. Each approach has its own advantages and disadvantages. However, there are not many literatures deliberate on the comparison of both approaches. This paper presents an overview of classification approaches in sentiment analysis. Various advantages and limitations of the sentiment classification approaches based on several criteria such as domain, classification type and accuracy are also discussed in this paper.

Keywords: Subjectivity sentiment analysis, sentiment classification, machine learning, lexicon-based.

1 Introduction

The rapid growth of content in the web has made a huge volume of information available. Users actively generate online information by expressing their thoughts, opinions or sentiments about all kinds of topics via blogs, forums and social network. This creates a great opportunity to understand the opinion of users by analyzing the rich content of opinionated data. Thus, it will be beneficial for both users and organizations. While some users generously express their thoughts and opinions about their points of interest in a fair environment, other users are able to seek for those reviews to gain helpful information in order to assist them in making decision such as in buying products or selecting the best services. Organizations also can get useful raw information; especially it is helpful in knowing better their customers and help to improve product quality or services performance. Sentiment analysis has gained much attention to researchers due to the overloaded online opinions and reviews. Unlike traditional text mining that concentrates on analysis of facts, sentiment analysis

© Springer Science+Business Media Singapore 2015
M.W. Berry et al. (Eds.): SCDS 2015, CCIS 545, pp. 43–53, 2015.
DOI: 10.1007/978-981-287-936-3_5

focuses on attitudes of users. Sentiment analysis is the process used to determine the attitude, opinion or emotion express by a person about a particular topic [1].

According to Liu [2], sentiment analysis is the study that analyzes people's sentiments, evaluations, attitudes and emotions in the favor of the entities such as products, services, organizations, individuals, issues, events, topics and their attributes. It is a recent research field, which uses advanced techniques for text mining, machine learning, information retrieval and NLP to process large amounts of nonstructural content generated by users [3]. Sentiment analysis has been used in many applications such as review related websites [4] and sub-component technology [5]. The most common application is the review-related websites and the most popular reviews are on products such as "Google Product Search" [6]. Review-related websites do not only restrict to product reviews, but also contains opinions about other reviews such as employers, political issues and so on. Recommender system is one of the most useful sentiment analyses for sub-component technology. It suggests the right items to particular users based on their explicit and implicit preferences by applying information filtering technologies [7]. Generally, the first task in sentiment analysis is to distinguish between objective and subjective statements. Subjective sentiments are words and phrase that being used to express mental and emotional states such as speculations, evaluations, sentiments and beliefs [8]. Meanwhile, objective content refers to factual information of the text sentence [6]. After the identification of subjective statement, polarity classification is then being performed. Polarity or also known as semantic orientation is important to identify the sentiment orientation whether it is a positive, negative or neutral sentiment [9]. Later, the orientation strength of the sentiment is identified. This paper aims to discuss available sentiment classification approaches. Section 2 reviews the related studies in sentiment analysis. Section 3 further explains each type of sentiment classification approaches. In section 4, we summarized the comparison between approaches in sentiment analysis. Finally in section 5, several points are brought into conclusion.

2   Related Works

Currently, many users actively express their opinions on the web. Opinions are central to human activities because they are key influencers of our behaviours [2]. It is a norm in life as we would like to know owners' opinions before we make decision. "What other people think" has always been an important piece of information for most of us during the decision making process [4]. Opinion can be regular or comparative [2]. Regular opinion describes about an entity with its aspects. This type of opinion is mean for single entity and not suitable for multiple entities. Meanwhile, a comparative opinion describes the relation of similarities or differences between several entities who shared several aspects. Thus, this type of opinion is suitable for multiple entities. In fact, we can classify opinion based on how it is being expressed in a text, either in explicit or implicit form [10]. Explicit opinion is directly expressed by users whereas implicit opinion is indirectly expressed by users. Sentiment analysis is classified into three main levels which are document-level, sentence-level and aspect/entity-level

[11]. The simplest form of sentiment analysis is document level. The aim is to classify a textual review which is given on a single topic as a positive or negative sentiment [12]. The whole review will be considered as either a positive or a negative review. A few researches exist at this level of analysis such as   Mouthami et al. [1], Moraes et al [12], Tang [13], Sharma & Dey [14] and Balamurali et al. [15]. While document-level assumes that a whole document express a single opinion, sentence-level analysis considers each sentence in documents as a separate unit. It means that every sentence in document carries out a single opinion with a certain orientation [5]. Positive or negative orientation means a sentence contains sentiment, while neutral orientation indicates that a sentence does not contains any sentiment. In sentence-level analysis, it is assumed that a sentence represents the opinion of just one author and a sentence hold the author's opinion about one topic only [16].

As most researchers focuses on a single level of analysis whether a document-level or a sentence-level, Maas et al.   [17] adopted their proposed vector-based approach model to both document-level and sentence-level sentiment analysis in online movie reviews domain. Meanwhile, aspect/entity-level focuses on the recognition of sentiment expressions and the aspects to which they refer [6]. It classifies sentiment with respect to the specific aspects of entities. In a review, people tend to talk about several entities that have many aspects and they have different opinion about each aspect of those entities. However, this may cause conflicting sentiments to assigning dual feeling to an entity or its attributes [18]. When a sentiment that lies between a positive and negative value, it might cause neutrality [19]. According to Medhat et al. [20], the process of sentiment analysis consists of several steps, which are sentiment identification, feature selection and sentiment classification. The first step is to identify a sentiment in a text. Basically in order to obtain the subjectivity content, pre-processing processes need to be done to obtain a clean data. Feature selection step will select appropriate text features in order to proceed with classification process. There are several feature selection methods available such as Information Gain, Mutual Information, Chi-square and Gain Ratio [21]. Finally, the sentiment polarity is obtained from the classification process. In sentiment analysis, several researchers have utilized available lexical database sources such as WordNet. WordNet is an English corpus which represents the meanings of words into synonym sets or known as synsets. Each synsets will have glosses that link the synsets according to other semantic relations such as hyponomy and holonomy [22]. Other lexical database is SentiWordNet which an improved version of WordNet lexical database developed by Esuli & Sebastiani [23]. They tagged all the synsets in WordNet according to three kinds of labels, which are positive, negative and objective. To date, the current official version of SentiWordNet is 3.0.

3   Sentiment Classification Approaches

Sentiment analysis classification approaches have evolved from keyword-based to concept-based approaches. Cambria et al. [24] grouped the existing approaches into four main categories, which are keyword spotting, lexical affinity, statistical method, and concept-based approach. Meanwhile, most researchers classify sentiment classification approaches into machine learning approaches and lexicon-based approaches [25], as shown in Fig. 1.

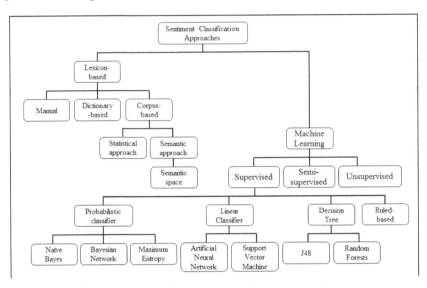

Fig. 1. Overview of sentiment classification approaches

3.1   Lexicon-Based Approach

Lexicon-based approach is unsupervised learning and does not require prior training in order to mine the data. Instead, it measures how far a sentiment word is inclined towards positive or negative [16; 26]. The aim is to identify the orientations of opinions expressed by users whether the opinion is positive, negative or neutral. Lexicon-based approach is further classified into manual approach, dictionary-based approach and corpus based approach. Manual approach requires people to code the lexicon by hand. It is very laborious, costly, can be only use on small corpora and usually a benchmarking procedure for automatic techniques [22]. Dictionary-based approach aims to identify sentiment sentences by utilizing synonyms and antonyms of available lexical databases such as WordNet. List of sentiment words with known orientation are manually collected. This approach expands the list of words by searching for their synonyms and antonyms in lexical databases. New found word is added iteratively to existing lists of words until no new words are found [5]. Corpus-based approach helps to solve the problem of identifying sentiment words with context specific orientation [20]. This approach depends on a seed of list words to find other of opinion words in a large corpus or explored the relation between the words [27]. It requires expensive manual annotation effort as it involves large corpus [28]. Corpus-based approach is further classified into statistical and semantic approach. The statistical approach aims to utilize statistics concept by finding the co-occurrences of words in order to obtain the sentiment polarity [29]. Meanwhile, semantic approach provides a semantic space to represent the terms in order to discover different kind of semantic relationship among terms. Words in semantic spaces are represented as high dimensional vectors and these vectors are obtained from statistical properties of the words' context [30]. Several recent researches in sentiment analysis using lexicon-based approaches are summarized in Table 1.

Table 1. Recent studies of lexicon-based approach in sentiment analysis.

Author	Techniques	Feature Selection	Level	Data/ Corpus	Result/ Performance	Limitation/ Future Work
[15]	Semantic space based on WordNet senses.	Word senses	-	Dataset by Ye et al. (2009): 600 positive and 591 negative reviews about travel destinations. Each review contains approximately 4 to 5 sentences with 80 to 85 average numbers of words	Accuracy of 91.12% with improvement of 6.2% compared to baseline method (word as feature).	To incorporate syntactical information with semantics. To explore cross lingual sentiment analysis.
[27]	General Inquirer, LBM, MPQA, K-means, ONMTF, Moodlens, CFMS and ESSA.	Word clusters	-	40216 tweets from Stanford Twitter Sentiment, collected between April 6, 2009 and June 25, 2009 via Twitter API. 3269 tweets from ObamaMcCain Debate on September 26, 2008.	The proposed framework of ESSA -opt performs better than baseline method, General Inquirer with improvement 21.4% and 17.87% for both datasets.	To study other emotion correlation information, to measure the sentiment orientation of social media posts.To utilize such models for real world applications.
[30]	HAL, COALS, RI, RRI and BEAGLE	Word clusters	Document level	Product and movie review dataset from Czech and English language.	RRI and COALS tend to perform consistently best for both languages.HAL and RI give no satisfactory results.	To investigate a combined model thatincludes the best performing semantic space for each cluster size.
[31]	MT, SMap, SProp1, SProp2 and SProp3.	Seed words	Document level	1000 positive and 1000 negative English movie reviews. 600 positive and 600 negative opinionated Dutch documents on 40 distinct topics crawled from Dutch review websites, forums, and blogs.	29% improvement from baseline in multi-language translation and 47% improvement in language specific semantic lexicon.	To validate the findings in other target language. To further optimize the seed sets used for the sentiment propagation process.
[32]	Naïve Bayes, Maximum Entropy (ME), SVM and Semantic analysis (WordNet).	Unigrams	Sentence level	19340 pre-labeled product review datasets from Twitter with of positive and negative polarity.	Semantic analysis (WordNet) achieves highest accuracy of 89.9%. Naïve Bayes with unigram model perform better thanSVM and ME with 88.2% accuracy	To improve feature vector related sentence identification process. To extend WordNet for reviews summarization.

3.2   Machine Learning Approach

Machine learning approach in sentiment analysis is basically a classification algorithm, which trains labelled documents over corpus in order to recognize features that will be used to classify the sentiment [33]. This approach can be supervised, semi-supervised or unsupervised. There are several types of supervised classifier, which are probabilistic, linear, rule based and decision tree classifiers [20]. As sentiment analysis is actually a typical classification task, it is most suitable to adopt supervised learning approach. Sentiment classification in machine learning approach consists of two main steps. The first step is to extract features from training data and convert into feature vectors. The second step is to train the classifier on feature vectors and application of the classifier to unseen instances [34]. Generally, two different set of data are required for the purpose of training and testing the algorithm. Training set needs to learn the diverse characteristic of documents with an automatic classifier, while a testing set will validate the performance of the classifier. Therefore, feature construction and classifier play a key role in classifying sentiment accurately. Meanwhile, according to Newel et al. [35], a machine learning system involves several steps starts with selecting a training data, data cleaning mechanism, labelling technique for the training data, feature vector selection and classification algorithm.

The research on sentiment analysis using machine learning approach is first done Pang et al. [36]. They conducted experiments by applying several supervised classifiers such as Naïve Bayes, Maximum Entropy and Support Vector Machine (SVM) on movie reviews dataset. They found that standard machine learning techniques outperform human-produced baseline. However, the performance of the classifiers does not perform well in classifying sentiment as compared to classify traditional text classification. The main reason may be that they generally follow traditional topical text classification approaches, where words in document are being set as a bag of words (BOW) concept. Traditional BOW does not capture word order information, syntactic structures and semantic relationships between words, which are essential attributes for sentiment analysis [37]. Many researchers applied machine learning approach for sentiment classification ranging from probabilistic classifiers such as

Table 2. Recent studies of machine learning approach in sentiment analysis.

Author	Techniques	Feature Selection	Level	Data/ Corpus	Result/ Performance	Limitation/ Future Work
[16]	Naïve Bayes and SVM	n-grams	Sentence level	1000 tweets data in Arabic language consisting of 500 positive and 500 negative tweets. The selected tweets hold only one opinion, not sarcastic, subjective and from different topics.	SVM is better than NB with 72.6% accuracy.	To improve the developed corpus by using fine-grained annotation and to add semantic features.
[35]	Naïve Bayes and SVM	Bag-of-words	Document level	Twitter datasets –30000 tweets randomly selected between January 23, 2011 and February 8, 2011 with keyword "Egyptian" and 12000 tweets randomly selected between November 5, 2012 and November 12, 2012 with keyword "Elections". Google Play on Android – 30000 positive and 30000 negative reviews.	SVM perform better than Naïve Bayes. SVM is more sensitive to outliers.	To consider varied capabilities of attacker in integrity attacks generated by sentiment analysis application.
[41]	SVM, ANN, Naïve Bayes, Random Forrest and J48.	Textual, categorical, grammatical and contextual features	-	1000 Norwegian financial internet news articles from publisher hegnar.no collected between February 4, 2013 and March 26, 2013.	J48 yield the highest performance, closely followed by Random Forrest.	To adopt the current algorithm in psychology-rich Norwegian financial domain.
[42]	Naïve Bayes, Binarized Multinomial Naïve Bayes (BMNB), Multinomial Naïve Bayes, SVM and J48.	Information gain, term frequency, term presence and part of speech	Sentence level	Total of 24000 sentences of multi-domain sentiment data (12 domains) from1000 positive and 1000 negative reviews from acquired from http://www.cs.jhu.edu/~mdredze/datasets / sentiment/	The BMNB clearly outperformed other classifiers in six data domain. SVM was ranked as the second best performing classifier in four out of twelve data domains. The best feature selection method is Information Gain.	-
[43]	Support Vector Machine (SVM), Naïve Bayes and K-nearest Neighbor	Information Gain, Principal Component Analysis, Relief-F, Gini Index, Uncertainty, Chi-square, SVM	-	2000 reviews from online Malay social media and blogs.1000 positive and 1000 negative reviews.	SVM perform the best with 87% accuracy. The worst performance is KNN classifier.	To develop Malay sentiment lexicons and to investigate the implementation of lexicon based approach.
[44]	SVM and Linear Regression	n-grams, sentiment bearing terms, rhetorical, length and positional features	Sentence level	News from English MOAT Research collection with 3584 sentences2697 sentences judged as objective and 887 judged as subjective polarity.	Proposed method outpre forms baseline method (OpinionFinder) with precision of 55.17% compared to 44.20% of baseline method.	To validate findings with other datasets and to study advanced ways in combining features and classifiers.
[45]	Naïve Bayes, SVM, Max imum Entropy and ensemble classifier.	Part of speech, special keyword, presence of negation, emoticon, positive and negative keywords, positive and negative hashtags.	-	1200 electronic product review from Twitter with 600 positive tweets and 600 negative tweets.	SVM and ME have equal accuracy of 90% compared to Naïve Bayes with 89.5% accuracy. Naïve Bayes has better precision	-
[46]	SVM	Bags of words using Vector Space Model	-	Benchmark data by Pang & Lee (2004). 2000 reviews of 54 different hotels data from popular travel destinations in India from website tripadvisor.com and yatra.com.	SVM with boosting algorithm showed the best accuracy of 93%.	To adopt boosting/ bagging approach in other classifiers and in other types of text such as social network.
[47]	Back-propagation artificial neural network (BPANN)	Information Gain	Document level	Benchmark data by Pang & Lee (2004). 501 positive and 501 negative user generated reviews for hotels. Utilized 3 sentimentlexicons: HM dataset – 1336 adjectives: 657 are positive and 679 are negative. General Inquirer dataset– 3596 adjectives, adverbs, nouns, and verbs: 1614 are positive and 1982 are negative. Opinion Lexicon – 2006 positive and 4783 negative words.	BPANN wi th Information Gain features gives the best accuracy of 95%. Opinion lexicon achieves of 89% and General Enquirer with 86% accuracy.	To adapt in other domain such as social network.

Naïve Bayes [38], Bayesian Network [39], and linear classifiers such as Artificial Neural Network [14] and Support Vector Machine [17; 40] and also with decision tree [41-42]. Besides, there are many researches done by comparing the capabilities among machine learning algorithms. Wang et al. [34] conducted a comparative assessment of ensemble learning techniques, which integrate different feature sets in order to enhance the classification accuracy. They choose five machine learning classifier which are Naïve Bayes, Maximum Entropy, Decision Tree, K Nearest Neighbor and SVM. They found that the ensemble technique with SVM produced the best accuracy. They are several recent researches in sentiment analysis using machine learning techniques are summarized in Table 2.

4   Comparison of Sentiment Classification Approaches

Each approach in sentiment classification has its own advantages and disadvantages as shown in Table 3. A lexicon-based approach is unsupervised and domain independent. It means one lexicon is built for all domains. It is an attractive advantage over machine learning methods, as this approach has a more robust performance across domains and texts [31]. On the other hand, machine learning approach is supervised and domain dependent. It is crucial to have a domain specific lexicon that is related to both the entities and their sentiment expressions in order to achieve a good sentiment analysis result [48]. Particularly, the same word could have completely opposite meanings or different sentiment strengths in different domains, thus it might affect the classification results. Machine learning approach requires data to be trained for domain specific polarity [16]. A key factor in determining the success of a machine learning approach is the quality and quantity of the training data. The more data, more consistent and noise-free it is, the better the results [49]. From the reviews, it can be seen that some researchers collect their own datasets from variety sources such as websites, forums and blogs. Meanwhile, there are numbers of researchers utilizing the available benchmark datasets developed by Pang and Lee in 2004 [46-47]. The dataset consists of 2000 movie reviews where it is associated with binary sentiment polarity label. Reviews with more than 3 star ratings considered as positive and reviews with less than 3 star ratings as negative. Reviews with 3 star ratings (neutral) are normally discarded [35; 46]. In lexicon-based approach, the maintenance of sentiment lexicons for different domains is an important issue [46]. Lexicon-based perform quicker than machine learning approach, but machine learning tends to be more accurate than lexicon-based approach [50].

Table 3. Comparison of lexicon-based and machine learning approach

Criteria	Lexicon-based	Machine learning
Domain	Independent	Dependent
Classification type	Unsupervised	Supervised
Prior training	No	Yes
Adaptive learning	No	Yes
Time to produce results	Fast	Slow
Maintenance	Require maintenance of corpus/corpora	Do not require maintenance
Accuracy	Less accurate than machine learning	Better than lexicon-based

5   Conclusion

People eagerly share and post content on the web expressing their points of view in an unrestricted way. The dimensionality and size of opinionated data are growing exponentially and turn out to be valuable sources to sentiment analysis. This paper discusses the classification approaches in sentiment analysis; lexicon-based and machine learning approaches. Both approaches have their own strengths in undergoing classification tasks. While lexicon-based approach is suitable for cross-domain application, machine learning approach is a domain specific. Machine learning approach is able to achieve better accuracy results as the context and semantic meaning is highly related. On the other hand, lexicon-based approach is a cross domain application and provides better coverage of lexicons. Thus it offers a wider scope of opinionated words to be classified accordingly. However, the issue of ambiguity of opinionated words needs to be considered. Some words behave in the same manner as they have the same meaning and polarity in all domains. Some words might behave inconsistently as they have different meaning and polarity in different domains.

References

1. Mouthami, K., Nirmala Devi, K., Murali Bhaskaran, V.: Sentiment analysis and classification based on textual reviews. In: 2013 International Conference on Information Communication and Embedded Systems (ICICES). IEEE (2013)
2. Liu, B.: Sentiment analysis and opinion mining. Synthesis Lectures on Human Language Technologies 5(1), 1–167 (2012)
3. Firmino Alves, A.L., et al.: A Comparison of SVM versus naive-bayes techniques for sentiment analysis in tweets: a case study with the 2013 FIFA confederations cup. In: Proceedings of the 20th Brazilian Symposium on Multimedia and the Web. ACM (2014)
4. Pang, B., Lee, L.: Opinion mining and sentiment analysis. Foundations and Trends in Information Retrieval 2(1–2), 1–135 (2008)
5. Jagtap, V.S., Pawar, K.: Analysis of different approaches to sentence-level sentiment classification. International Journal of Scientific Engineering and Technology 2, 164–170 (2013). ISSN: 2277-1581
6. Feldman, R.: Techniques and applications for sentiment analysis. Communications of the ACM 56(4), 82–89 (2013)
7. Zhang, Z., et al.: A hybrid fuzzy-based personalized recommender system for telecom products/services. Information Sciences 235, 117–129 (2013)
8. Akkaya, C., Wiebe, J., Mihalcea, R.: Subjectivity word sense disambiguation. In: Proceedings of the 2009 Conference on Empirical Methods in Natural Language Processing, vol. 1. Association for Computational Linguistics (2009)
9. Martın-Wanton, T., et al.: Word sense disambiguation in opinion mining: Pros and cons. Special Issue: Natural Language Processing and its Applications 119, 358 (2010)
10. Gryc, W., Moilanen, K.: Leveraging textual sentiment analysis with social network modelling. From Text to Political Positions: Text Analysis Across Disciplines 55, 47 (2014)
11. Kansal, H., Toshniwal, D.: Aspect based Summarization of Context Dependent Opinion Words. Procedia Computer Science 35, 166–175 (2014)

12. Moraes, R., Valiati, J.F., Neto, W.P.G.: Document-level sentiment classification: An empirical comparison between SVM and ANN. Expert Systems with Applications 40(2), 621–633 (2013)
13. Tang, D.: Sentiment-specific representation learning for document-level sentiment analysis. In: Proceedings of the Eighth ACM International Conference on Web Search and Data Mining. ACM (2015)
14. Sharma, A., Dey, S.: A document-level sentiment analysis approach using artificial neural network and sentiment lexicons. ACM SIGAPP Applied Computing Review 12(4), 67–75 (2012)
15. Balamurali, A.R., Joshi, A., Bhattacharyya, P.: Robust sense-based sentiment classification. In: Proceedings of the 2nd Workshop on Computational Approaches to Subjectivity and Sentiment Analysis. Association for Computational Linguistics (2011)
16. Shoukry, A., Rafea, A.: Sentence-level Arabic sentiment analysis. In: 2012 International Conference on Collaboration Technologies and Systems (CTS). IEEE (2012)
17. Maas, A.L., et al.: Learning word vectors for sentiment analysis. In: Proceedings of the 49th Annual Meeting of the Association for Computational Linguistics: Human Language Technologies, vol. 1. Association for Computational Linguistics (2011)
18. Boiy, E., Moens, M.-F.: A machine learning approach to sentiment analysis in multilingual Web texts. Information Retrieval 12(5), 526–558 (2009)
19. Khan, K., Baharudin, B.B., Khan, A.: Mining opinion from text documents: A survey. In: 3rd IEEE International Conference on Digital Ecosystems and Technologies, DEST 2009. IEEE (2009)
20. Medhat, W., Hassan, A., Korashy, H.: Sentiment analysis algorithms and applications: A survey. Ain Shams Engineering Journal 5(4), 1093–1113 (2014)
21. Abdul-Rahman, S., et al.: Exploring feature selection and support vector machine in text categorization. In: 2013 IEEE 16th International Conference on Computational Science and Engineering (CSE). IEEE (2013)
22. Boia, M.: Context Sensitive Sentiment Analysis. Thesis director, pp. 1–8 (2012)
23. Esuli, A., Sebastiani, F.: Sentiwordnet: A publicly available lexical resource for opinion mining. In: Proceedings of LREC, vol. 6 (2006)
24. Cambria, E., et al.: New avenues in opinion mining and sentiment analysis. IEEE Intelligent Systems 28(2), 15–21 (2013)
25. Saif, H., Fernandez, M., He, Y., Alani, H.: SentiCircles for contextual and conceptual semantic sentiment analysis of twitter. In: Presutti, V., d'Amato, C., Gandon, F., d'Aquin, M., Staab, S., Tordai, A. (eds.) ESWC 2014. LNCS, vol. 8465, pp. 83–98. Springer, Heidelberg (2014)
26. Vinodhini, G., Chandrasekaran, R.M.: Sentiment analysis and opinion mining: a survey. International Journal 2(6) (2012)
27. Hu, X., et al.: Unsupervised sentiment analysis with emotional signals. In: Proceedings of the 22nd International Conference on World Wide Web. International World Wide Web Conferences Steering Committee (2013)
28. He, Y., Zhou, D.: Self-training from labeled features for sentiment analysis. Information Processing & Management 47(4), 606–616 (2011)
29. Turney, P.D.: Thumbs up or thumbs down?: semantic orientation applied to unsupervised classification of reviews. In: Proceedings of the 40th Annual Meeting on Association for Computational Linguistics. Association for Computational Linguistics (2002)
30. Habernal, I., Brychcín, T.: Semantic spaces for sentiment analysis. In: Habernal, I., Brychcín, T. (eds.) TSD 2013. LNCS (LNAI), vol. 8082, pp. 484–491. Springer, Heidelberg (2013)

31. Hogenboom, A., et al.: Multi-lingual support for lexicon-based sentiment analysis guided by semantics. Decision Support Systems 62, 43–53 (2014)
32. Gautam, G., Yadav, D.: Sentiment analysis of twitter data using machine learning approaches and semantic analysis. In: 2014 Seventh International Conference on Contemporary Computing (IC3). IEEE (2014)
33. Giannakopoulos, G., et al.: Representation models for text classification: a comparative analysis over three Web document types. In: Proceedings of the 2nd International Conference on Web Intelligence, Mining and Semantics. ACM (2012)
34. Wang, G., et al.: Sentiment classification: The contribution of ensemble learning. Decision Support Systems 57, 77–93 (2014)
35. Newell, A., et al.: On the practicality of integrity attacks on document-level sentiment analysis. In: Proceedings of the 2014 Workshop on Artificial Intelligent and Security Workshop. ACM (2014)
36. Pang, B., Lee, L., Vaithyanathan, S.: Thumbs up?: sentiment classification using machine learning techniques. In: Proceedings of the ACL 2002 Conference on Empirical Methods in Natural Language Processing, vol. 10. Association for Computational Linguistics (2002)
37. Xia, R., Zong, C., Li, S.: Ensemble of feature sets and classification algorithms for sentiment classification. Information Sciences 181(6), 1138–1152 (2011)
38. Almatrafi, O., Parack, S., Chavan, B.: Application of location-based sentiment analysis using Twitter for identifying trends towards Indian general elections 2014. In: Proceedings of the 9th International Conference on Ubiquitous Information Management and Communication. ACM (2015)
39. He, Y.: A bayesian modeling approach to multi-dimensional sentiment distributions prediction. In: Proceedings of the First International Workshop on Issues of Sentiment Discovery and Opinion Mining. ACM (2012)
40. Das, A., Björn, G.: Sentimantics: conceptual spaces for lexical sentiment polarity representation with contextuality. In: Proceedings of the 3rd Workshop in Computational Approaches to Subjectivity and Sentiment Analysis. Association for Computational Linguistics (2012)
41. Njolstad, P.C.S., et al.: Evaluating feature sets and classifiers for sentiment analysis of financial news. In: 2014 IEEE/WIC/ACM International Joint Conferences on Web Intelligence (WI) and Intelligent Agent Technologies (IAT), vol. 2. IEEE (2014)
42. Saad, F.: Baseline evaluation: an empirical study of the performance of machine learning algorithms in short snippet sentiment analysis. In: Proceedings of the 14th International Conference on Knowledge Technologies and Data-driven Business. ACM (2014)
43. Alsaffar, A., Omar, N.: Study on feature selection and machine learning algorithms for Malay sentiment classification. In: 2014 International Conference on Information Technology and Multimedia (ICIMU). IEEE (2014)
44. Chenlo, J.M., Losada, D.E.: A machine learning approach for subjectivity classification based on positional and discourse features. In: Lupu, M., Kanoulas, E., Loizides, F. (eds.) IRFC 2013. LNCS, vol. 8201, pp. 17–28. Springer, Heidelberg (2013)
45. Neethu, M.S., Rajasree, R.: Sentiment analysis in twitter using machine learning techniques. In: 2013 Fourth International Conference on Computing, Communications and Networking Technologies (ICCCNT). IEEE (2013)
46. Sharma, A., Dey, S.: A boosted svm based sentiment analysis approach for online opinionated text. In: Proceedings of the 2013 Research in Adaptive and Convergent Systems. ACM (2013)

47. Sharma, A., Dey, S.: A document-level sentiment analysis approach using artificial neural network and sentiment lexicons. ACM SIGAPP Applied Computing Review 12(4), 67–75 (2012)
48. Mudinas, A., Zhang, D., Levene, M.: Combining lexicon and learning based approaches for concept-level sentiment analysis. In: Proceedings of the First International Workshop on Issues of Sentiment Discovery and Opinion Mining. ACM (2012)
49. Devitt, A., Ahmad, K.: Is there a language of sentiment? An analysis of lexical resources for sentiment analysis. Language Resources and Evaluation 47(2), 475–511 (2013)
50. Dang, Y., Zhang, Y., Chen, H.: A lexicon-enhanced method for sentiment classification: An experiment on online product reviews. IEEE Intelligent Systems 25(4), 46–53 (2010)

Comparisons of ADABOOST, KNN, SVM and Logistic Regression in Classification of Imbalanced Dataset

Hezlin Aryani Abd Rahman[1], Yap Bee Wah[1], Haibo He[2], and Awang Bulgiba[3]

[1] Faculty of Computer and Mathematical Sciences,
Universiti Teknologi MARA Malaysia
40450 Shah Alam
[2] Department of Electrical, Computer and Biomedical Engineering
University of Rhode Island, Kingston, RI 02881, USA
[3] Julius Centre University of Malaya
Department of Social and Preventive Medicine
Faculty of Medicine, University of Malaya,
50603 Kuala Lumpur, Malaysia
{hezlin,beewah}@tmsk.uitm.edu.my,
he@ele.uri.edu,
awang@um.edu.my

Abstract. Data mining classification techniques are affected by the presence of imbalances between classes of a response variable. The difficulty in handling the imbalanced data issue has led to an influx of methods, either resolving the imbalance issue at data or algorithmic level. The R programming language is one of the many tools available for data mining. This paper compares some classification algorithms in R for an imbalanced medical data set. The classifiers ADABOOST, KNN, SVM-RBF and logistic regression were applied to the original, random oversampling and undersampling data sets. Results show that ADABOOST, KNN and SVM-RBF exhibits over-fitting when applied to the original dataset. No overfitting occurs for the random oversampling dataset where by SVM-RBF has the highest accuracy (Training: 91.5%, Testing: 90.6%), sensitivity (Training :91.0%, Testing: 91.0%), specificity (Training: 92.0%,Testing: 90.2%) and precision (Training:91.9%, Testing 90.5%) for training and testing data set. For random undersampling, no overfitting occurs only for ADABOOST and logistic regression. Logistic regression is the most stable classifier exhibiting consistent training an testing results.

Keywords: imbalanced data set, ADABOOST, KNN, SVM, logistic regression.

1 Introduction

The spread of *Big Data* has invaded the health-care sector when Google announced its venture with Calico in 2013 [1]. Google, being one of the many

© Springer Science+Business Media Singapore 2015
M.W. Berry et al. (Eds.): SCDS 2015, CCIS 545, pp. 54–64, 2015.
DOI: 10.1007/978-981-287-936-3_6

multi-billion-dollar companies looking to exploit this convergence of data, technology and health care. The move highlighted the uprising interest in the potential for massive data to promote innovation and efficiency in the health care sector. Among one of the many challenges of *Big Data* is the *imbalanced data* captured and stored in the medical databases. Almost all of the real-world data sets are naturally *imbalanced* to a certain degree [2]. This *imbalanced* nature are commonly encountered in a wide range of application in various field of study such as prediction of customer churn [3] and fraud detection [4] or identifying potential breast cancer, hepatitis, hyperthyroid, and cardiovascular disease [5], and analyzing alzheimer disease through neuroimaging [6].

In critical area such as the health and medical field, study related to classification and predictive models, especially in prediction of deaths or diseases tend be more costly when the predictive model is inaccurate due to the imbalanced issue. Nonetheless, the standard machine learning algorithm in data mining tools would not perform well in the presence of imbalanced data sets [7]. Thus, for a binary response variable, when the event of interest is underrepresented (the minority class), the majority class tends to hinders the classification accuracy of standard machine learning algorithm.

In recent years, many methods have been developed to improve the classification of imbalanced data. Estabrook & Japkowicz [10] proposed a mixture-of-experts or hybrid approach that is induced the classifier after over-sampling or under-sampling the data with different over or undersampling rate. Although, these new hybrid methods were reported to be effective in handling the imbalanced data sets, nonetheless there were other issues such as over-fitting, potential data loss and over-complexity of developed models. Bekkar et. al. [8] provided a framework of approaches in handling IDS whereby the approaches are classified into five main categories: sampling methods, ensemble learning, cost-sensitive learning, feature selection methods and algorithm modification. He & Garcia [2] reviewed the state-of-the-art technologies, and the current assessment metrics used to evaluate learning performance under the imbalanced learning scenario. Recently, Bekkar et. al. [8] compiled a comprehensive overview of methods in handling imbalanced data set from the data mining practitioner point of view. The evaluation level is another perspective that the data scientist should explore in countering the issues of handling imbalanced data sets [2, 7, 9].

Although complex methods have been developed, many data mining practitioners still prefer to handle this issue at data level [9, 16], using sampling approach. This is due to the ease of application and simplicity of the random oversampling or undersampling technique. Many machine learning algorithms are available in R, an open source programming language. This paper illustrates the hybrid approach using sampling techniques and induced with a classifier to improve classification of an imbalanced medical dataset. The classifiers considered are logistic regression, ADABOOST, Support Vector Machine (SVM) and k-Nearest Neighbors (KNN). The analyses were carried out using R. This paper is organized as follows: Section 2 reviews some methods in handling IDS. Section 3 covers the methodology and the results are presented in Section 4. Finally, Section 5 concludes the paper.

2   Literature Review

2.1   Machine Learning Techniques

Many machine learning techniques have been developed in the past few decades. Over the years, machine learning has grown and merge to become a broad discipline in producing many fundamental theories of learning processes and algorithms. The main theoretical advances have catered supervised learning problems, where the classifier is estimated using the labeled observations drawn from the population. Among examples of these type of supervised learning is the logistic regression, decision tree C5.0 and neural networks. Recent developments involve support vector machine (SVM) and k-nearest neighbors (KNN) approach.

Logistic Regression. Logistic Regression is the most popular classifier in medical research [13–16]. This is due the fact that logistic regression, since it was proposed as an alternative to the ordinary least squares regression in 1970s, delivers the important information for doctors and medical practitioners such as the odds ratio, relative risk and predictions of dichotomous outcomes. In addition, the ordinary least squares methods requires strict statistical assumption's such as linearity and normality [16].

In order to model a dichotomous (0,1) outcome, the first issue was that dichotomous in nature do not follow a linear trend. Secondly, the errors are neither normally distributed nor constant [17]. Therefore, Logistic regression solves these problems, through the application of the logit transformation to the dependent variable. In other words, the logistic model predicts the logit of Y from X, which is the natural logarithm (ln) of odds of Y, and odds ratios is given by $\exp(\beta)$. The following equation shows the simple logistic regression model:

$$logit(Y) = naturallog(odds) = ln(\frac{\pi}{1-\pi}) = \alpha + \beta X \tag{1}$$

The R syntax used to obtain a logistic regression model is as follows:

```
#formula to set Logistic Regression Model
card.formula <-die~gender+age_grp+comorbid+surgery+creopen
+afib+winf+euroscore
#Run GLM formula
card.glm <-glm(card.formula, family=binomial (link="logit"),
data=data.train)
```

Boosting with AdaBoost. Adaboost or Adaptive Boosting is an iterative boosting approach to improve the classification of the minority class, especially when dealing with IDS. At the initial stage, the Adaboost algorithm will allocate variant weights to each observations. After a few iteration, the weight impose on the misclassified observations will increase, and vice versa, the correctly classified will have lesser weights. The weights on the observations are the indicators as to which class the observation belongs to, thus lower the misclassification of the observations while tremendously improve the performance of the classifiers at the same time. The process of Adaboost is explained by Bauer & Kohavi [21] and Freund, Schapire & Hill [22]. The following is the syntax in R:

```
#Run adabag package
data.adaboost <- boosting(card.formula, data=data.train,
boos= TRUE, mfinal = 10, control = (minsplit=0))
```

K-Nearest Neighbors. *KNN* is one of the informed sampling technique which is quite popular among data scientists. The idea of KNN as an informed sampling technique was highlighted by [2]. The strategy of KNN is to classify the observations are based on the surrounding *neighbors*, where k in *KNN* refers to the number of *neighbors* to be considered in the sampling process.

For example, let us consider x as the observation to be classified. Let us assume that $k=3$, thus we consider that the sign of x be determined by three of its neighbors with the nearest distance; x_1, x_2 and x_3.

When investigated, x_1 and x_2 are labeled as negative and only x_3 is a positive. Therefore, x is also labeled as a negative case based on the majority of the neighbors, whereby two out of three of the neighbors are negative. The distance used to calculate the closest *neighbors* are based on the Euclidean distance formula [2].

In order to prevent attributes with initially large ranges from outweighing attributes with initially smaller ranges i.e binary attributes, the Euclidean distance equation are normalized. Min-max normalization, is an example of a popular normalizing technique that is used to transform a value v of a numeric attribute A to v' in the range [0, 1] by computing using the following formula:

$$v' = \frac{v - min_A}{max_A - min_A} \tag{2}$$

where min_A and max_A are the minimum and maximum values of attribute A [23]. The R syntax is as follows:

```
##Run kknn package for train dataset
data.kknn <- kknn(card.formula, data.train, data.train, k=1,
distance = 1,kernel = "optimal")
data.kknn
```

Support Vector Machine. The Support Vector Machine (SVM) is the current popular data mining classification algorithm and is claimed to be the most versatile classifier by data mining community. Many studies have used SVM in handling IDS and reported good results [24–27].

The architecture of SVMs uses the general category of kernel methods. These kernels depend on the distribution of the data mapped through some dot-products, whereby these dot-products are then transformed into kernel functions which computes a dot-product in high dimensional feature space. As an efficient classifier, SVM is able generate non-linear decision boundaries using methods designed for linear classifiers. In addition, the kernel functions used in SVM allows the researchers to apply the classifier to any data sets which have no obvious representation in terms of patterns. The strengths of SVM has made it popular in scientific and medical studies [25, 27].

SVM was used to classify and recognize human movement patterns by [26], while [24] performed an experimentation of different kernels functions on SVM and concluded that SVM is very successful when applied to IDS with negative instances heavily outnumbering the positive instances.

The classification of SVM is achieved by realizing a linear or non-linear separation surface in the input space. In SVM classification, the separating function can be expressed as a linear combination of kernels associated with the Support Vectors as expressed in equation 3.

$$f(x) = \sum_{x_j \epsilon S} \alpha_j y_j K(x_j, x) + b \qquad (3)$$

Remark 1. Figure 1 illustrates the flexibility of the kernels in SVM. In this example, the linear kernel already shows good separation between the red and blue dots. However, using the polynomial kernel, the separation between the two-colored are more prominent especially when using polynomial kernel with 1 or 2 degrees.

The following is the R syntax used in this study:

```
##Run svm for train dataset
svm.model <- svm(card.formula, data = data.train, kernel_type=2)
svm.test <- predict(svm.model, data.test)
```

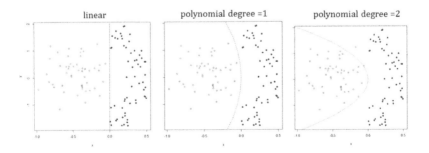

Fig. 1. Different degree of polynomial kernels effects on SVM.

2.2   Oversampling and Undersampling

The oversampling and undersampling technique are considered as the basic sampling techniques in handling imbalanced data sets [8]. The idea of oversampling, also known as up-sampling, is to duplicate the minority class observations until the number of observations balanced the majority class. In this study we use the random oversampling whereby the proportion of the minority and majority class were balanced out by duplicating the observations in the minority class in a random manner [8, 11].

In contrast, the undersampling or known as down-sampling, excludes the observations from the majority class to balance with the number of available observations in the minority class. The undersampling technique selected in this study is the random undersampling. The random undersampling randomly excludes the observations in the majority class until balance is achieved between both majority and minority class. As for the comparison in terms of performance between oversampling and undersampling, some studies concluded that undersampling outperformed oversampling in terms of accuracy of decision tree C4.5 or CART [8, 12].

Many studies uses oversampling and undersampling with decision tree as the classifier [19, 20]. Thus, the aim of this paper is to compare the performance of these sampling techniques with different classifiers; Logistic Regression, SVM and KNN, to decision tree C5.0. The following code is used to perform random oversampling and undersampling in R.

```
#balanced data set with over-sampling (ROS)
data.over <- ovun.sample(card.formula, data=data.dt,
p=0.5, seed=s,method="over")$data
table(data.over$die)
# balanced data set with under-sampling (RUS)
data.under <- ovun.sample(card.formula, data=data.dt,
p=0.5, seed=s,method="under")$data
table(data.under$die)
```

3   Method

3.1   Data Set

This study only focuses on the classification of the binary (or two classes) imbalanced problems. The positive instances belong to the minority class and the negative instances belong to the majority class. The data set used in this study is the cardiac surgery data set obtained from a local medical center in Kuala Lumpur. The data set contains cases from a study on prediction of survival of cardiac surgery patients. The data set consists of 8 independent variables (7 categorical and 1 interval type data) and one binary outcome variable. There were a total of 4976 observations, whereby 4767 observations (95.8%) is the patients were still Alive after surgery (majority class) and only 209 (4.2%) had 'Died' after surgery. This strongly shows that the cardiac surgery data set obtained is an imbalanced data set. The eight independent variables were: gender (f,m); Age Group (18-40, 40-60, above 60); Comorbidities (Hypertension, Diabetes, Both, None); Surgery Type (CABG only, CABG and Valve Surgery, Others); Chest Reopen (Yes, No); Atrial Fibrillation (Yes, No), Wound Infection (Yes, No); EUROScore. There were no problems of imbalanced data for the categorical predictors.

3.2   Methods

The first part of analysis is to identify the effect of imbalance towards the classification performance of logistic regression as classifier. Thus, the data set will be modeled using Logistic Regression and the performance will be analyzed. The second part of the analysis is to determine the effect when different approaches in handling imbalanced data set is applied accordingly, namely, oversampling and undersampling. The evaluation measures used in this study are the Accuracy, Sensitivity, Specificity and Precision.

Sensitivity and Specificity are important measures especially in the context of medical data analysis, whereby, these measures are usually used in order to assess the effectiveness of a clinical test in detecting a given disease. Unlike classification accuracy, these two measures are not affected by imbalanced data as sensitivity and specificity identifies the proportions of correct classification for Positive and Negative event. Meanwhile, precision measures how precise is a classifier in predicting the outcome of interest correctly that is, how much of positive predictions are correct (Number of positive observations divided by number of observation assigned to the positive class). The technique used to evaluate a model is very much as important as the technique used in building the model itself [11].

First the original imbalanced cardiac surgery dataset was partitioned into 70% training and 30% testing samples. The four classifiers were applied to the training samples and evaluated using the testing sample. Next, we balanced the data using random oversampling (ROS) and random undersampling (RUS). Then, the ROS was partitioned into 70% training and 30% testing samples, and the four classifiers were then applied. We repeat the same procedure for the RUS data set. The experiments were performed using the classification algorithms available in different packages in R (glm, libSVM, adabag, kknn, ROSE).

4   Results and Discussions

First, we apply logistic regression to the original data set. The original (imbalanced data) yields a very high classification accuracy and specificity rate (96.1% and 99.5%). However, the sensitivity rate was very poor (19.2%), indicating that the *imbalanceness* of the data set has an effect towards the performance of the logistic regression model. Then, we apply logistic regression to the ROS and RUS samples. Although the accuracy decreased, the testing sensitivity improves for

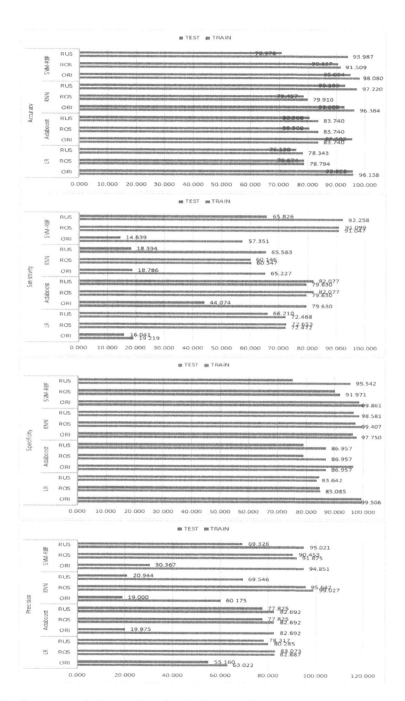

Fig. 2. Training and Testing Results for Logistic Regression, Adaboost, KNN and SVM-RBF (%).

Table 1. Logistic Regression, Adaboost, KNN and SVM-RBF Results (%)

| Technique | Evaluation | Dataset | | | | | |
| | | Original | | ROS | | RUS | |
		Training	Testing	Training	Testing	Training	Testing
Logistic Regression	Accuracy	96.138	95.858	78.794	78.674	78.343	76.130
	Sensitivity	19.219	16.041	72.472	72.653	72.468	66.210
	Specificity	99.506	99.386	85.085	84.865	83.642	84.512
	Precision	63.022	55.160	82.887	83.073	80.285	78.317
Adaboost	Accuracy	83.740	95.582	83.740	80.508	83.740	80.508
	Sensitivity	79.630	44.074	79.630	82.077	79.630	82.077
	Specificity	86.957	96.587	86.957	79.113	86.957	79.113
	Precision	82.692	19.975	82.692	77.825	82.692	77.825
KNN	Accuracy	96.384	93.009	79.910	78.457	97.220	93.189
	Sensitivity	65.227	18.786	60.347	60.146	65.563	18.394
	Specificity	97.750	96.264	99.407	97.193	98.581	96.608
	Precision	60.175	19.000	99.027	95.642	69.546	20.944
SVM-RBF	Accuracy	98.080	95.004	**91.509**	**90.637**	93.987	70.976
	Sensitivity	57.351	14.639	**91.047**	**91.099**	92.258	65.826
	Specificity	99.861	98.543	**91.971**	**90.166**	95.542	75.230
	Precision	94.851	30.367	**91.875**	**90.453**	95.021	69.326

the ROS (72.7%) and RUS (66.2%) data sets. In addition, for ROS dataset, all variables were found to be significant except comorbidities.

Next, we apply ADABOOST, KNN and SVM approach to the original, ROS and RUS data sets. The results in Table 1 shows over-fitting occurs when SVM-RBF, ADABOOST, and KNN were applied to the original imbalanced and RUS data set. This could be due to the small number of positive instances in these data sets. The over-fitting problem did not occur in the ROS data set.

Overall, the performance of all classifier for the training and testing samples results are more consistent for the ROS data set as shown in Figure 2. All the methods seemed to improve when applied with ROS with SVM-RBF having the highest accuracy, sensitivity, specificity and precision. Unfortunately, for RUS, the over-fitting issue arises for KNN and SVM-RBF methods.

The results of this study shows that a combination of ROS with classifier such as Logistic Regression, Adaboost, KNN and SVM can improve the classification of imbalanced data set. The logistic regression is also more stable and less prone to over-fit the data as can be seen by the consistent results in both the training and testing for original, ROS and RUS samples. The stability of Logistic Regression as a classifier emerged in this experiment.

5   Conclusions

Sampling approaches are always easier to implement in improving predictions of the minority case of a two-class classification problem. Both sampling techniques presented in this paper comes with their own advantages and disadvantages.

The application in this study shows there were no over-fitting problem when classifier were applied to the random oversampling data set. This experiment results favors random oversampling. Work is in progress which involves a simulation study to investigate different sampling approaches for imbalanced data set. It is also important to note that the classifiers performance depend on sample size and data quality. All data sets should be cleaned and imbalanced problems in categorical predictor (or features) should be determined so as to obtain a good predictive model with results that can be generalized. Future studies can apply other sampling techniques such as Tomek-Links, SMOTE, Border-line SMOTE, One-sided Selection, Neighborhood Cleaning Rule and Bootstrap-based Oversampling.

Acknowledgments. We thank the Research Management Institute (RMI) Universiti Teknologi MARA and the Ministry of Higher Education (MOHE) Malaysia for the funding of this research under the Malaysian Fundamental Research Grant, 600- RMI/FRGS 5/3 (16/2012).

References

1. Ward, A.: Interest in Healthcare 'Big Data' Grows. FT.com. ProQuest. Web (2014), February 10, 2015
2. He, H., Garcia, E.A.: Learning from Imbalanced Data. IEEE Transactions on Knowledge and Data Engineering 21(9), 1263–1284 (2009)
3. Nikulin, V., McLachlan, G.J.: Classification of imbalanced marketing data with balanced random sets. In: JLMR Workshop and Conference Proceedings, vol. 7, pp. 89–100 (2009)
4. Ogwueleka, F.: Data Mining Application in Credit Card Fraud Detection System. J. Eng. Sci. Technol. 6(3), 311–322 (2011)
5. Mena, L., Gonzalez, J.A.: Machine learning for imbalanced datasets: application in medical diagnostic. In: Proceedings of the Nineteenth International Florida Artificial Intelligence Research Society Conference (FLAIRS 2006), pp. 574–579 (2006)
6. Dubey, R., Zhou, J., Wang, Y., Thompson, P.M., Ye, J., and Alzheimer's Disease Neuroimaging Initiative.: Analysis of sampling techniques for imbalanced data: An n = 648 ADNI study. Analysis of sampling techniques for imbalanced data: An n = 648 ADNI study. NeuroImage 87, 220–241 (2014)
7. Weiss, G.M.: Foundations of imbalanced learning. In: He, H., Ma, Y. (eds.) Imbalanced Learning, Foundations, Algorithms, Applications, 1st edn., pp. 13–42. Wiley and IEEE Press, New Jersey (2013)
8. Bekkar, M., Alitouche, T.A.: Imbalanced Data Learning Approaches Review. International Journal of Data Mining and Knowledge Management Process (IJDKP) 3(4), 15–33 (2013)
9. Bekkar, M., Djemaa, H.K., Alitouche, T.A.: Evaluation Measures for Models Assessment over Imbalanced Data Sets. Journal of Information Engineering and Applications 3(10), 27–39 (2013)
10. Estabrooks, A., Japkowicz, N.: A mixture-of-experts framework for learning from unbalanced data sets. In: Hoffmann, F., Adams, N., Fisher, D., Guimarães, G., Hand, D.J. (eds.) IDA 2001. LNCS, vol. 2189, pp. 34–43. Springer, Heidelberg (2001)

11. Japkowicz, N.: Learning from imbalanced data sets: A comparison of various strategies. In: AAAI Workshop on Learning from Imbalanced Data Sets, pp. 1–5 (2000)
12. Chawla, N.V., Bowyer, K.W., Hall, L.O., Kegelmeyer, W.P.: SMOTE: Synthetic Minority Over-sampling Technique. Journal of Artificial Intelligence Research 16, 321–357 (2002)
13. Chang, Y.: Boosting SVM classifiers with logistic regression, pp. 1–16 (1995). See www.stat.sinica.edu.tw/library/c_tec_rep/2003 (2003)
14. Everitt, B.S., Hothorn, T.: Logistic regression and generalised linear models: blood screening, womens role in society, and colonic polyps. In: A Handbook of Statistical Analyses Using R, 1st edn., pp. 97–112. Taylor and Francis Group (LLC), London (2006)
15. Jiang, X., El-Kareh, R., Ohno-Machado, L.: Improving predictions in imbalanced data using pairwise expanded logistic regression. In: Annual Symposium Proceedings / AMIA Symposium. AMIA Symposium, 2011, pp. 625–634 (2011)
16. Sathian, B.: Reporting dichotomous data using Logistic Regression in Medical Research: The scenario in developing countries. Nepal Journal of Epidemiology 1(4), 111–113 (2011)
17. Peng, C.-Y.J., Lee, K.L., Ingersoll, G.M.: An Introduction to Logistic Regression Analysis and Reporting. The Journal of Educational Research 96(1) (2010)
18. Kubat, M., Matwin, S.: Addressing the curse of imbalanced training sets: one-sided selection. In: Proceedings of the Fourteenth International Conference on Machine Learning, vol. 4, pp. 179–186 (1997)
19. Batista, G., Prati, R.C., Monard, M.C.: A study of the behavior of several methods for balancing machine learning training data. ACM SIGKDD Explorations Newsletter 6(1), 20 (2004)
20. Estabrooks, A., Jo, T., Japkowicz, N.: A Multiple Resampling Method for Learning from Imbalanced Data Sets. Computational Intelligence 20(1), 18–36 (2004)
21. Bauer, E., Kohavi, R.: An empirical comparison of voting classification algorithms: Bagging, boosting, and variants. Machine Learning 139, 105–139 (1999)
22. Freund, Y., Schapire, R.E., Hill, M.: Experiments with a new boosting algorithm. In: 13th International Conference on Machine Learning (1996)
23. Han, J., Kamber, M.: Data Mining Concepts and Techniques (A. Stephan, Ed.), 2nd edn., vol. 40. Morgan Kaufmann Publishers Inc and Elsevier Inc., San Francisco (2006)
24. Akbani, R., Kwek, S., Japkowicz, N.: Applying support vector machines to imbalanced datasets. In: Boulicaut, J.-F., Esposito, F., Giannotti, F., Pedreschi, D. (eds.) ECML 2004. LNCS (LNAI), vol. 3201, pp. 39–50. Springer, Heidelberg (2004)
25. Auria, L., Moro, R.A.: Support Vector Machines (SVM) as a Technique for Solvency Analysis, pp. 1–16. Discussion Papers of Deutsches Institute of Wirtschaftsforschung, Berlin (2008)
26. Schuldt, C., Laptev, I., Caputo, B.: Recognizing human actions: a local SVM approach. Pattern Recognition, 3–7 (2004)
27. Jiang, X., El-Kareh, R., Ohno-Machado, L.: Improving predictions in imbalanced data using pairwise expanded logistic regression. In: Annual Symposium Proceedings (AMIA Symposium), pp. 625–634 (2011)
28. Yap, B.W., Rahman, H.A.A., He, H., Bulgiba, A.: Handling imbalanced dataset using SVM and k-NN approach. In: Simposium Kebangsaan Sains Matematik (SKSM22) (2014) (in Press)

Finding Significant Factors on World Ranking of e-Governments by Feature Selection Methods over KPIs

Simon Fong[1], Yan Zhuang[1], Huilong Luo[1], Kexing Liu[1], and Gia Kim[2]

[1] Deparment of Computer and Information Science, University of Macau, Macau SAR
[2] Leaders' Partner, Melbourne, Australia
{ccfong,syz,mb25509,mb45462}@umac.mo,
gia.kim@leaderspartner.org

Abstract. Computing significant factors quantitatively is an imperative task in understanding the underlying reasons that contribute to a final outcome. In this paper, a case of e-Government ranking is studied by attempting to find the significance of each KPIs which leads to resultant rank of a country. Significant factors in this context are inferred as some degrees of relations between the input variables (which are the KPIs in this case) and the final outcome (the rank). In the past, significant factors were either acquired as first-hand information via direct questioning from users' satisfaction survey or qualitative inference; typical question is 'You are satisfied with a particular e-Government service' by applying a multi-level Likert scale. Respondents answered by choosing one of the following: Strongly agree, Agree, Neutral, Disagree, and Strongly Disagree. The replies are then counted and studied using traditional statistical methods. In this paper, an alternative method by feature selection in data mining is proposed which computes quantitatively the relative importance of each KPI with respective to the predicted class, the rank. The main advantage of feature selection by data mining (FSDM) method is that it considers the cross-dependencies of the variables and how they contribute as a whole predictive model to a particular predicted outcome. In contrast, classical significant factor analysis such as correlogram tells only the strength of correlation between an individual pair of factor and outcome. Another advantage of using data mining method over simple statistic is that the inferred predictive model could be used as a predictor and/or what-if decision simulator; given some values of KPIs a corresponding rank could be guesstimated. A case study of computing significant factors in terms of KPIs that lead to the world rank in from the data of UN e-Government Survey 2010, is presented.

Keywords: e-Government, Ranking, Analytics, Feature Selection, Data Mining, Principle of Component Analysis.

1 Introduction

KPI (key performance indicator) method now is widely used for performance evaluation field. KPI method is the combination of MBO (Management By Objective) and

© Springer Science+Business Media Singapore 2015
M.W. Berry et al. (Eds.): SCDS 2015, CCIS 545, pp. 65–73, 2015.
DOI: 10.1007/978-981-287-936-3_7

Pareto's Law ("20/80" laws) which is a decomposition of the subject's strategic objective, analysis and it summarizes the critical success factors to support the subject's strategic objective, and then it extracts the key performance indicators. Based on this background, the core idea we can conclude as the 80% of the subject's performance can be hold and lead by 20% of the key indicators, all the performance evaluation work should focus on those 20% of key indicators [1].

One drawback however with this KPI methodology is that there allows no reverse in finding out how these KPIs contribute to the final rank, though the relative importance of KPI comes from expert judgement. In this paper, we attempt to infer the relative importance of KPIs by finding out how much they contribute to a resultant rank of an e-Government from the feedbacks collected from survey.

Some domestic and international academic communities or commercial organizations have already engaged in this research field, such as Accenture consulting firm, who used Service Maturity and Delivery Maturity those two key categories label to indicate the government network (URL: http://www.aceenture.com). Meanwhile, World Markets Research Center and Brown University, they embark on one research by using 22 refined indicators for measuring 2288 government websites over 196 countries and regions (URL: http://www.worldmarketsanalysis.com). Most papers in the literatures [2]-[6] cover extensively about qualitative research of e-Government, like the evaluation framework research case study and some factors affecting the performance evaluation analysis. A typical study which was the Economic Research Center of Peking University, who was commissioned by the National Advisory Committee of Information Technology. They published their research results in the October 2003, which are based on 257 prefecture-level cities' government website of China (URL: http://gov.cn). Besides, most of the current researches are in theoretical study, which focus on the strategic choice and potential problems in the context of management. Some others explored the problems from different angles to expound the design of e-Government performance evaluation index system.

All these studies point to a fact that finding significance of each KPI that lead to a world rank is important. In the view that little previous works have studied the use of data mining model in inferring the significance factors quantitatively with respect to world rank of e-Government, this paper aims at shedding some light on the evaluation aspect.

2   Empirical Data Analysis

This section is on about an empirical data analysis to be carried out for finding the significance of each factor pertaining to the world ranks of e-Governments. The KPIs are listed in Table 1 are estimated.

The experiment consists of two parts. Firstly the KPIs are computed as a correlogram by Pearson correlation algorithm which represents a classical approach for evaluating the importance of each variable on the target. Then a popular feature selection approach, known as Principle Component Analysis (PCA) is applied for comparing with the results of the correlogram. It is interesting to observe whether there would be difference in terms of the significances of KPIs obtained from between mere correlation and PCA of FSDM.

The second part of the experiment is on applying a collection of feature selection algorithms as FSDM in evaluating the KPIs.

2.1   The Dataset

A set of EU e-Government performance records is downloaded from Eurostat (URL: http://ec.europa.eu) – a Directorate-General of the European Commission located in Luxembourg. Its main responsibilities are to provide statistical information to the institutions of the European Union (EU).

A streamlined dataset consists of three groups of KPIs - ICT, G2B and G2C is collected from 28 EU countries. Thirteen representative KPIs are selected for this experimentation from the three groups. The year is particularly chosen to be 2010 which marks the year of European sovereign debt crisis. The same dataset was used in the 6th edition of the annual "Waseda World e-Government Ranking" which has been carried out since 2005. Over the past six years, the Waseda e-Government research team has surveyed the developments and observed the trends in the e-government arena. In particular these 13 KPIs from the three groups ICT, G2B and G2C were discussed in the Waseda e-Government 2010 report as new trends of e-Government development deduced from the survey. The final rankings and evaluation information for those 28 EU countries are taken from the report of United Nations e-Government Survey 2010, Leveraging e-government at a time of financial and economic crisis. (URL: http://www.unpan.org/egovkb/global_reports/08report.htm). The evaluation results in ranks and indices are appended to the KPI dataset. The ranks are used as a target class for inducing a prediction model which will be used to support feature selection for estimating the importance of each attribute.

Table 1. KPIs used in the experiment.

Nation	ICT			G2B					G2C				
	Households with broadband access	Percentage of the ICT personnel on total employment	Percentage of the ICT sector on GDP	Enterprises using the Internet for interaction with public authorities	Broadband and connectivity - enterprises	Enterprises using the Internet for submitting a proposal in a public electronic tender system to public authorities	Integration of internal processes	Share of enterprises/ turnover on e-commerce	Broadband and connectivity - Individuals(V1)	E-Government usage by individuals(V2)	Individuals frequently using the Internet(V3)	Individuals using the Internet for taxing part in online consultations or voting(V4)	Individuals ICT capability(V5)
EU-28	61.00	2.57	4.14	76	85	13	21	14	63	31	62	8	31
Belgium	70.00	3.00	5.00	77	90	10	40	18	73	32	68	4	39
Bulgaria	26.00	2.00	5.00	64	62	8	11	2	39	15	43	3	21
Czech Republic	54.00	3.00	5.00	89	86	11	21	19	61	17	54	3	31
Denmark	80.00	3.00	5.00	92	87	11	29	17	84	72	84	11	36
Germany	75.00	2.00	4.00	67	89	13	29	18	75	37	68	10	41
Estonia	64.00	3.00	5.00	80	88	17	7	11	68	48	63	6	23
Ireland	58.00	3.00	4.00	87	87	32	20	24	61	27	61	3	36
Greece	41.00	3.00	4.00	77	81	11	36	17.5	98	13	47	5	25
Spain	57.00	2.00	4.00	67	95	9	22	11	54	32	54	10	30
France	66.00	3.00	4.00	78	93	16	24	13	71	36	66	10	29
Croatia	49.00	2.00	4.00	63	78	16	15	9	48	16	53	7	25
Italy	49.00	2.00	4.00	84	84	10	22	5	44	17	54	6	20
Cyprus	51.00	2.00	4.00	74	85	3	17	1	44	22	53	3	24
Latvia	53.00	2.00	3.00	72	68	12	8	7	57	31	60	4	22
Lithuania	54.00	2.00	2.00	95	81	31	11	14	54	22	53	6	17
Luxembourg	70.00	2.00	3.00	90	87	14	21	15	88	55	82	13	37
Hungary	52.00	4.00	6.00	71	79	15	8	16	57	28	62	3	24
Malta	69.00	3.00	3.00	77	92	13	18	15	59	28	59	9	24
Netherlands	80.00	3.00	3.00	95	91	9	22	14	89	59	83	7	36
Austria	64.00	2.00	3.00	75	82	15	25	13	88	39	63	11	38
Poland	57.00	2.00	3.00	89	69	14	11	8	54	21	47	2	28
Portugal	50.00	2.00	4.00	75	85	20	26	12	45	23	48	5	15
Romania	23.00	1.00	3.00	50	52	16	19	4	31	7	32	2	25
Slovenia	62.00	2.00	4.00	88	88	11	21	10	62	40	58	8	30
Slovakia	49.00	3.00	5.00	88	78	7	17	11	67	35	61	3	29
Finland	76.00	4.00	5.00	96	96	13	28	18	82	54	80	19	48
Sweden	83.00	4.00	6.00	90	91	19	35	18	88	62	81	14	38
United Kingdom	58.70	3.00	6.00	66	88	10	6	16	79	40	78	8	38

2.2   Feature Selection Algorithm

As the core functions of FSDM, Feature Selection (FS) algorithms play a central role in calculating the "worthiness" of each attribute as significant factor. By observing from their results, we can estimate how much each KPI as well as the KPI group contributes to the final world ranking of the e-Government. Quantitatively the relative importance of each KPI is inferred as a numeric indicator, usually normalized to be in the range of [0, 1]. The computing software for FSDM used is called Weka which is a popular data mining software platform for benchmarking machine learning algorithms. The seven FS algorithms as FSDM that are used in the experiment are briefly presented as follow:

(1) Correlation: The algorithm is implemented as a Weka filter function called CfsSubsetEval. Each attribute is evaluated according to its predictive power, so that permuted subsets of the attributes can be found. The worthiness of each subset is computed from the collective predictive power of each attribute contained in the group. At the same time, redundancy between the corresponding attribute and its subgroup is considered too. Out of the qualified subgroups, those in which the attributes are greatly correlated with the target class and not so correlated to the attributes from the other groups are chosen as the result [7].

(2) Chi Square: This function implements Pearson's chi-squared test, which is also commonly known as likelihood ratio. It works by evaluating the goodness of fit over the relationship between the attribute value and the target class. The goodness of fit is defined by chi-squared statistic relating to the class [8].

(3) Info Gain: This function is called InfoGainAttributeEval in Weka. Information gain is known as the amount of chaos reduction in terms of information entropy that is contained in a dataset. It finds a group of attributes by recursively partitioning the dataset as per the information gain calculated from the divided data partition [9]. It aims at finding a set of attributes that carry the greatest information gain pertaining to the target class.

(4) ReliefFAttributeEval: This function [10] recurrently samples a sample instance of the data and considers the nearest neighbor instances of the data by their relations between the attribute and target class.

(5) Significant Attribute: It is known as SignificanceAttributeEval in Weka. The function [11] calculates the worthiness of each attribute by taking into account of the probability of occurrences of the associations between: classes-to-attribute, and attribute-to-classes. The probabilities are then checked against the highest probabilities which represent the most significant associations. If the new one has a higher significance, the current best attribute is replaced with it. Working from class to class, the attributes whose significance values occupy the top of the list are chosen.

(6) Symmetry Uncertainty: The algorithm [12] is implemented in Weka as a function called SymmetricalUncert-AttributeEval. It measures the pairwise mutual information between a pair of data objects. The mutual information is taken as the worthiness based on the symmetrical uncertainty observed during the process.

(7) Principal component: Principal component analysis (PCA) [13] measures the correlations between pairs of attributes-to-attributes and attributes-to-class. It then

transforms the attributes orthogonally to a set of eigenvectors which correlate the attribute values linearly to the uncorrelated variables. By knowing the property of principal components, PCA shrinks the dimensionality by taking the strongest eigen-vector(s) and eliminates those that have 20% or more of variance from the initial data.

2.3   The Experimental Results

The collection of feature-selection methods are applied over the set of e-Government KPIs (as in Table 1) each of which represents a features that describes the relation to the rank. The feature selection results in terms of the worthiness of each feature (KPI) are tabulated in Table 2. The feature selection results are all normalized via Min-Max conversion, to be numeric scores in the range of [0, 1]. The scores that are corres-ponding to different attribute numbers are calculated by averaging over each individ-ual score which is known as a rank or ranking information generated by the feature selection methods. The group score is averaged sum of the normalized score of each feature under the same group. The scores are charted as a bar chart with error-bars indicating the standard deviations of the corresponding influence factor of each KPI.

By observing over the results in Table 2 as well as visually in Figure 1, it is found that the top four contributing KPIs to their values of the ranks are:

1. Broadband and connectivity – individuals
2. E-Government usage by individuals
3. Individuals frequently using the Internet
4. Households with broadband access

They possess significance scores in average, 0.95, 0.95, 0.92 and 0.91 respectively. Relatively these four KPIs have the narrowest error-bars in comparison to the rest of the KPIs. For group-KPIs, the ranked groups are G2C, G2B and ICT, at averaged group scores of 0.860190992, 0.66592497, and 0.365380368 respectively.

2.4   Comparison with Other Methods

In order to validate the results obtained from FSDM, the same KPI dataset is subject to correlation analysis and PCA for generating and comparing the results with those ob-tained from FSDM. A correlogram is generated by R which is statistical software, and it is shown in Figure 2. It is a correlation matrix with the most correlated variables

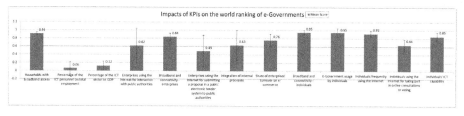

Fig. 1. Bar chart of averaged KPIs significance scores.

Table 2. The significance scores of KPIs by FSDM.

Group / KPI	ICT			GBB					GIC			
PS algorithm	Households with broadband access	Percentage of the ICT personnel on total employment	Percentage of the ICT sector on GDP	Enterprises using the Internet for interaction with public authorities	Broadband and connectivity enterprises	Enterprises using the Internet for submitting a proposal in a public electronic tender system to public authorities	Integration of internal processes	Share of enterprises' turnover on e-commerce	Broadband and connectivity individuals	E-Government usage by individuals	Individuals frequently using the Internet	Individuals using the Internet for taking part in online consultations or voting	Individuals' ICT capability
CorrelationAttribute	0.890838786	0.379966686	0.178762841	0.139797739	0.730371366	0	0.389302297	0.643664485	1	0.933819155	0.9985981	0.75594684	0.990026093
Chi-Squared Attribute	0.892887143	0	0.182601579	0.840225564	0.84022564	0.62699248	0.787593985	0.787593985	0.892857143	1	0.84022564	0.67368211	0.74982406
Info Gain	0.941241408	0	0.16859378	0.514534016	0.00319080	0.734275036	0.871314001	0.863584227	0.941241408	1	0.90719089	0.050178253	0.83900921
ReliefFAttribute	0.892857143	0	0.182601579	0.840225564	0.84022564	0.62699248	0.787593985	0.787593985	0.892857143	1	0.84022564	0.67508421	0.734962406
Significant Attribute	0.975460123	0	0.20228589	0.99825609	0.983122609	0.69558521	0.944783276	0.941717791	0.975460123		0.98322609	0.819018405	0.92847853
Symmetry Uncertainty	0.87285237	0	0	0	0.74366074	0	0	0.511560694	1	0.74566074	1	0.647398844	0.862716763
Total Score	5.46608715	0.379966686	0.734792786	3.70204552	5.02193076	2.92817954	3.78261553	4.5153251508	5.70264017	5.67493895	5.50612750	3.62649276	5.09141601
Mean Score	0.91101452	0.063681114	0.12240546	0.61713425	0.83712812	0.48812842	0.39105552	0.75053861	0.95042669	0.94058649	0.91768712	0.60748794	0.84833572
Standard Deviation	0.03890467	0.15467757	0.08916439	0.42809078	0.08931093		0.39195552	0.369407781	0.15908974	0.10912019	0.06684711	0.14221025	0.10202755
Group mean score	0.36508568			0.0870497					0.86690992				
Group Std. Deviation	0.09928624			0.2677645					0.09262188				

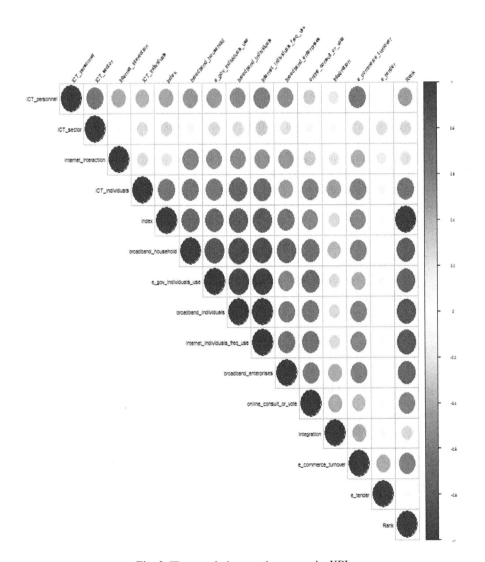

Fig. 2. The correlation matrix among the KPIs.

highlighted in strong colour and dot size in a diagonal data table. In this graph, correlation coefficients is coloured as dots in the matrix cells according to the coefficient values. By scanning through the most right hand column labelled as Rank, a series of dots of various sizes in maroon colour is displayed. The dots indicate the correlation-strength between each of the KPIs and the Rank. Index and Rank are most strongly correlated, though this relation is not useful because the world rank of an e-Government comes directly from the index which is some form of final score about the quality of the e-Government. The top four contributing KPIs in this case, according to the correlogram in Figure 2, are ranked and listed as follow by the correlation strength:

1. Individuals frequently using the Internet
2. Broadband and connectivity - individuals
3. Households with broadband access
4. E-Government usage by individuals

The correlation coefficients are of values, -0.79160, -0.78733, -0.76786, and -0.73186 respectively. It is noted that the four top significant KPIs selected by correlation method are the same as those by FSDM; however the orders of significance are different. It may be due to the fact that the correlation method calculates the significance coefficient on per-pair basis between an individual KPI and the class. FSDM takes into account of dependencies of features (KPIs) for inferring a model mapping to a class.

PCA is used to replicate the results. PCA is designed to choose a subset of features with strongest predictive powers. The goal of PCA is to transform the initial features into a new set that are sufficient to explain the variation in the data. The new features come in the form of eigenvectors that correspond to a linear combination of the original features called principal components. It is shown in Figure 3 that the first principal component occupies over 50% of variance in the dataset; thereby it strongly indicates that this linear combination of features should be selected as a significant set of features. And these features which are the e-Government KPIs are listed as:

1. Broadband and connectivity – individuals
2. E-Government usage by individuals
3. Individuals frequently using the Internet
4. Households with broadband access

The PCA values are sorted as 0.357, 0.356, 0.353 and 0.336. The results by PCA tally with those by FSDM. The contributing powers of these features are visualized as a factor map in Figure 3.

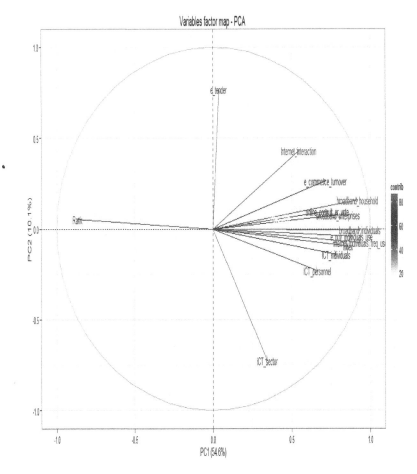

Fig. 3. Variable Factor Map.

3   Conclusions

Given a rank of an e-Government, it would be useful to know some insights at least in the form of how much does each underlying factor contribute to the resultant rank. As a case of world ranking of e-Government, the KPIs are taken as variables whose extents of significances are quantitatively computed as a result of contributing factors for the rank. In this paper, an analytic called FSDM is studied for estimating the significance factors. The FSDM method is compared with some classical significant factor analysis such as correlogram and PCA for validating its efficacy. The results generated by FSDM and PCA are almost identical, while correlogram selects the same top influencing variables but they come with different impacts. The slight differences of the results are probably due to the lack of cross-dependency when it comes to computing pairwise correlation. It is also noted that another advantage of using data mining method over simple statistic is that the inferred predictive model could be used as a predictor and/or what-if decision simulator; given some values of KPIs a corresponding rank could be guesstimated. Further investigation along the research direction of applying induced prediction model warrants our future work. So that users can test and predict the world rank of an e-Government by inputting the available KPIs values into the prediction model.

Acknowledgement. The authors of this paper would like to thank Research and Development Administrative Office of the University of Macau, for the funding support of this project which is called "Building Sustainable Knowledge Networks through Online Communities" with the project code MYRG2015-00024-FST.

References

1. Bender, P.: Mathematical Modeling of trie 20/80 Rule: Theory and Practice. Journal of Business Logistics 2(2), 139–157 (1981)
2. Bhatti, M.I., Awan, H.M., Razaq, Z.: The key performance indicators (KPIs) and their impact on overall organizational performance. Quality & Quantity 48(6), 3127–3143 (2014)
3. Gupta, M.P., Jana, D.: E-government evaluation: A framework and cage study. Government Information Quarterly 20(4), 365–387 (2003)
4. Akmana, I., Yazicib, A., Mishraa, A., Arifoglu, A.: E—Government: A global view and an empirical evaluation of some attributes of citizens. Government Information Quarterly 22(2), 239–257 (2005)
5. Montagna, J.M.: A framework for the assessment and analysis of electronic government proposal. Electronic Commerce Research and Applications 4(3), 204–219 (2005)
6. Steyaert, J.: Measuring the Performance of Electronic Government Services. Information and Management (41), 369–375 (2004)
7. Hall, M.A.: Correlation-based Feature Subset Selection for Machine Learning. University of Waikato, PhD Thesis, Hamilton, New Zealand (1998)
8. Bidgoli, A.M., Parsa, M.N.: A Hybrid Feature Selection by Resampling, Chi-squared and Consistency Evaluation Techniques. World Academy of Science, Engineering and Technology 68, 276–285 (2012)
9. Mukras, R., Wiratunga, N., Lothian, R., Chakraborti, S., Harper, D.: Information gain feature selection for ordinal text classification using probability Re-distribution. In: Proceedings of the Textlink workshop at IJCAI 2007, pp. 1–10 (2007)
10. Robnik-Sikonja, M., Kononenko, I.: An adaptation of Relief for attribute estimation in regression. In: Fourteenth International Conference on Machine Learning, pp. 296–304 (1997)
11. Ahmad, A., Dey, L.: A feature selection technique for classificatory analysis. Pattern Recognition Letters 26(1), 43–56 (2005)
12. Kannan, S., Ramaraj, N.: A novel hybrid feature selection via Symmetrical Uncertainty ranking based local memetic search algorithm. Journal of Knowledge-Based Systems 23(6), 580–585 (2010)
13. Jolliffe, I.T.: Principal Component Analysis. Springer Series in Statistics, 2nd edn. (2002). ISBN 0-387-95

Part II
Fuzzy Computing

Possibility Vague Soft Expert Set Theory and Its Application in Decision Making

Ganeshsree Selvachandran[1] and Abdul Razak Salleh[2]

[1] Department of Actuarial Science and Applied Statistics,
Faculty of Business and Information Science, UCSI University,
Jalan Menara Gading, 56000 Cheras, Kuala Lumpur, Malaysia
[2] School of Mathematical Sciences, Faculty of Science and Technology,
Universiti Kebangsaan Malaysia, 43600 UKM Bangi, Selangor DE, Malaysia
ganeshsree86@yahoo.com, aras@ukm.edu.my

Abstract. In this paper, we aim to extend the notion of classical soft expert sets to possibility vague soft expert sets by applying the theory of soft expert sets to possibility vague soft sets. The complement, union, intersection, AND and OR operations as well as some related concepts pertaining to this notion are defined. The algebraic properties such as the De Morgan's laws and the relevant laws of possibility vague soft expert sets are studied and subsequently proved. Lastly, this concept are applied to a decision making problem and its effectiveness is demonstrated using a hypothetical example.

Keywords: Vague soft set, soft expert set, fuzzy soft expert set, soft set.

1 Introduction

Soft set theory was first proposed by Molodtsov in 1999 ([1]) as a general mathematical tool for dealing with uncertainties, imprecision and vagueness that cannot be handled using classical mathematical tools. Since its inception, many generalizations of this theory have been introduced. Maji et al. ([2]) established the concept of fuzzy soft sets as an extension to the notion of classical soft sets and studied its properties. Majumdar and Samanta ([3]) introduced and studied the concept of generalized fuzzy soft sets where a degree is attached with the parameterization of fuzzy sets while defining a fuzzy soft set. Alkhazaleh et al. ([4]) then introduced the theory of soft multisets and fuzzy soft multisets ([5]) as a generalization of soft set theory. They also defined the concepts of fuzzy parameterized interval-valued fuzzy soft sets ([6]), possibility fuzzy soft sets ([7]), generalized interval-valued fuzzy soft sets ([8]) and gave their applications in decision making and medical diagnosis. Furthermore, Alkhazaleh and Salleh introduced the concept of soft expert sets ([9]) and subsequently the concept of fuzzy soft expert sets ([10]). Hassan and Alhazaymeh introduced the theory of vague soft multisets ([11]), interval-valued vague soft sets ([12]), generalized vague soft sets ([13]), possibility vague soft sets ([14]) and vague soft expert set theory ([15]) and studied the application of these theories in decision making and medical diagnosis

© Springer Science+Business Media Singapore 2015
M.W. Berry et al. (Eds.): SCDS 2015, CCIS 545, pp. 77–87, 2015.
DOI: 10.1007/978-981-287-936-3_8

problems. This trend is continued in this paper through the establishment of the notion of possibility vague soft expert sets (denoted as PVSES from now on). PVSESs can better handle the elements of imprecision and uncertainty that arises in assigning a suitable membership function to an element compared to the other generalizations of soft expert sets such as fuzzy soft expert sets, vague soft expert sets and possibility vague soft sets. The PVSES model is also significantly more advantageous compared to possibility fuzzy soft sets as it has the added advantage of allowing the users to know the opinion of all the experts in one model without the need for any operations. Moreover, even after performing any operations, the users can still know the opinion of all the experts. In line with this, the purpose of this paper is to extend the classical soft expert set model to the PVSES model and thereby establish a new generalization of the soft expert set model called the possibility vague soft expert set (PVSES).

2   Preliminaries

In this section, we present some relevant background knowledge pertaining to the concepts used in this paper.

Definition 2.1 ([1]). A pair (F, A) is called a *soft set* over U, where F is a mapping given by $F: A \rightarrow P(U)$. In other words, a soft set over U is a parameterized family of subsets of the universe U. For $\varepsilon \in A$, $F(\varepsilon)$ may be considered as the set of ε-elements of the soft set (F, A) or as the ε-approximate elements of the soft set.

Definition 2.2 ([17]). Let X be a space of points (objects) with a generic element of X denoted by x. A *vague set* V in X is characterized by a truth-membership function $t_V : X \rightarrow [0, 1]$ and a false-membership function $f_V : X \rightarrow [0, 1]$. The value $t_V(x)$ is a lower bound on the grade of membership of x derived from the evidence for x and $f_V(x)$ is a lower bound on the negation of x derived from the evidence against x. The values $t_V(x)$ and $f_V(x)$ both associate a real number in the interval $[0, 1]$ with each point in X, where $t_V(x) + f_V(x) \leq 1$. This approach bounds the grade of membership of x to a subinterval $[t_V(x), 1 - f_V(x)]$ of $[0, 1]$. Hence a vague set is a form of fuzzy set, albeit a more accurate form of fuzzy set.

Definition 2.3 ([17]). The *complement* of a vague set A, denoted by A' and is defined as given below:

$$t_{A'}(x) = f_A(x) \qquad \text{and} \qquad 1 - f_{A'}(x) = 1 - t_A(x).$$

Definition 2.4 ([16]). A pair $(\hat{F}, A)$ is called a *vague soft set* over U where $\hat{F}$ is a mapping given by $\hat{F}: A \rightarrow V(U)$ and $V(U)$ is the power set of vague sets over U. In other words, a vague soft set over U is a parametrized family of vague sets of the universe U. Every set $\hat{F}(e)$ for all $e \in A$, from this family may be considered as the set of e-approximate elements of the vague soft set $(\hat{F}, A)$. Hence the vague soft set $(\hat{F}, A)$ can be viewed as consisting of a collection of approximations of the following form:

$$\left(\hat{F}, A\right) = \left\{\hat{F}(x_i): i = 1, 2, 3, \dots\right\} = \left\{\frac{\left[t_{\hat{F}(e_i)}(x_i), 1 - f_{\hat{F}(e_i)}(x_i)\right]}{x_i} : i = 1, 2, 3, \dots\right\}$$

for all $e \in A$ and for all $x \in U$.

3   Possibility Vague Soft Expert Sets

In this section, the notion of possibility vague soft expert sets are established and the properties of this concept are then studied and discussed.

From now on, let U be universal set of elements, E be a set of parameters, X be a set of experts (agents), Q be a set of opinions, $Z = E \times X \times Q$ and $A \subseteq Z$.

Definition 3.1. Let $U = \{u_1, u_2, u_3, \dots, u_n\}$ be a universal set of elements, $E = \{e_1, e_2, e_3, \dots, e_m\}$ be a universal set of parameters, $X = \{x_1, x_2, x_3, \dots, x_i\}$ be a set of experts (agents) and $Q = \{1 = agree, 0 = disagree\}$ be a set of opinions. Let $Z = \{E \times X \times Q\}$ and $A \subseteq Z$. Then the pair (U, Z) is called a soft universe. Let $\tilde{F} : Z \to I^U$ and μ be a vague subset of Z defined as $\mu : Z \to I^U$, where I^U denotes the collection of all vague subsets of U. Suppose $\tilde{F}_\mu : Z \to I^U \times I^U$ be a function defined as:

$$\tilde{F}_\mu(z) = \left(\tilde{F}(z)(u_i), \mu(z)(u_i)\right), \qquad \forall u_i \in U.$$

Then $\tilde{F}_\mu$ is called a *possibility vague soft expert set* (denoted as PVSES for simplicity) over the soft universe (U, Z).

For each $z_i \in Z$, $\tilde{F}_\mu(z_i) = \left(\tilde{F}(z_i)(u_i), \mu(z_i)(u_i)\right)$ where $\tilde{F}(z_i)$ represents the degree of belongingness of the elements of U in $\tilde{F}(z_i)$ and $\mu(z_i)$ represents the degree of possibility of such belongingness. Hence $\tilde{F}_\mu(z_i)$ can be written as:

$$\tilde{F}_\mu(z_i) = \left\{\left(\frac{u_i}{\tilde{F}(z_i)(u_i)}\right), \mu(z_i)(u_i)\right\}, \quad \text{for } i = 1, 2, 3, \dots$$

where $\tilde{F}(z_i)(u_i) = \left[t_{\tilde{F}(z_i)}(u_i), 1 - f_{\tilde{F}(z_i)}(u_i)\right]$, with $t_{\tilde{F}(z_i)}(u_i)$ and $f_{\tilde{F}(z_i)}(u_i)$ representing the truth membership function and false membership function of each of the elements $u_i \in U$ respectively.

Often the PVSES $\left(\tilde{F}_\mu, Z\right)$ can be written simply as $\tilde{F}_\mu$. If $A \subseteq Z$, it is also possible to have a PVSES $\left(\tilde{F}_\mu, A\right)$.

Example 3.2. Let $U = \{u_1, u_2, u_3\}$ be a set of elements, $E = \{e_1, e_2\}$ be a set of decision parameters, where e_i $(i = 1, 2, 3)$ denotes the parameters $E = \{e_1 = beautiful, e_2 = cheap\}$ and $X = \{x_1, x_2\}$ be a set of experts. Suppose that $\tilde{F}_\mu : Z \to I^U \times I^U$ is a function defined as follows:

$$\tilde{F}_\mu(e_1, x_1, 1) = \left\{\left(\frac{u_1}{[0.8, 0.9]}, 0.3\right), \left(\frac{u_2}{[0.4, 0.5]}, 0.2\right), \left(\frac{u_3}{[0.6, 0.8]}, 0.3\right)\right\}, \dots,$$

$$\tilde{F}_\mu(e_2, x_2, 0) = \left\{\left(\frac{u_1}{[0.3, 0.5]}, 0.55\right), \left(\frac{u_2}{[0.7, 0.9]}, 1\right), \left(\frac{u_3}{[1, 1]}, 0.99\right)\right\}.$$

Then we can view the possibility vague soft expert set $(\tilde{F}_\mu, Z)$ as consisting of the following collection of approximations:

$$(\tilde{F}_\mu, Z) = \left\{(e_1, x_1, 1) = \left\{\left(\frac{u_1}{[0.8, 0.9]}, 0.3\right), \left(\frac{u_2}{[0.4, 0.5]}, 0.2\right), \left(\frac{u_3}{[0.6, 0.8]}, 0.3\right)\right\}\right\}, \dots,$$

$$\left\{(e_2, x_2, 0) = \left\{\left(\frac{u_1}{[0.3, 0.5]}, 0.55\right), \left(\frac{u_2}{[0.7, 0.9]}, 1\right), \left(\frac{u_3}{[1, 1]}, 0.99\right)\right\}\right\}.$$

Then $(\tilde{F}_\mu, Z)$ is a possibility vague soft expert set over the soft universe (U, Z).

Definition 3.3. Let $(\tilde{F}_\mu, A)$ and $(\tilde{G}_\delta, B)$ be PVSESs over a soft universe (U, Z). Then $(\tilde{F}_\mu, A)$ is said to be a *possibility vague soft expert subset* (PVSE subset) of $(\tilde{G}_\delta, B)$ if $A \subseteq B$ and for all $\varepsilon \in A$, the following conditions are satisfied:

(i) $\mu(\varepsilon)$ is a vague subset of $\delta(\varepsilon)$,
(ii) $\tilde{F}(\varepsilon)$ is a vague subset of $\tilde{G}(\varepsilon)$.

This relationship is denoted as $(\tilde{F}_\mu, A) \subseteq (\tilde{G}_\delta, B)$. In this case, $(\tilde{G}_\delta, B)$ is called a *possibility vague soft expert superset* (PVSE superset) of $(\tilde{F}_\mu, A)$.

Definition 3.4. Let $(\tilde{F}_\mu, A)$ and $(\tilde{G}_\delta, B)$ be PVSESs over a soft universe (U, Z). Then $(\tilde{F}_\mu, A)$ and $(\tilde{G}_\delta, B)$ are said to be *equal* if for all $\varepsilon \in E$, the following conditions are satisfied:

(i) $\mu(\varepsilon)$ is equal to $\delta(\varepsilon)$,
(ii) $\tilde{F}(\varepsilon)$ is equal to $\tilde{G}(\varepsilon)$.

In other words, $(\tilde{F}_\mu, A) = (\tilde{G}_\delta, B)$ if $(\tilde{F}_\mu, A)$ is a PVSE subset of $(\tilde{G}_\delta, B)$ and $(\tilde{G}_\delta, B)$ is a PVSE subset of $(\tilde{F}_\mu, A)$.

Definition 3.5. A PVSES $(\tilde{F}_\mu, A)$ is said to be a *possibility null vague soft expert set*, denoted by $(\tilde{\emptyset}_\mu, A)$ and defined as:

$$(\tilde{\emptyset}_\mu, A) = \left(\tilde{F}(\alpha), \mu(\alpha)\right), \qquad \forall \alpha \in Z,$$

where $\tilde{F}(\alpha) = 0$, that is $t_{\tilde{F}(\alpha)} = 0$ and $f_{\tilde{F}(\alpha)} = 1$ and $\mu(\alpha) = 0$ for all $\alpha \in Z$.

Definition 3.6. A PVSES $(\tilde{F}_\mu, A)$ is said to be a *possibility absolute vague soft expert set*, denoted by $(\tilde{F}_\mu, A)_{Abs}$ and defined as:

$$(\tilde{F}_\mu, A)_{Abs} = \left(\tilde{F}(\alpha), \mu(\alpha)\right), \qquad \forall \alpha \in Z,$$

where $\tilde{F}(\alpha) = 1$, that is $t_{\tilde{F}(\alpha)} = 1$ and $f_{\tilde{F}(\alpha)} = 0$ and $\mu(\alpha) = 1$ for all $\alpha \in Z$.

Definition 3.7. Let $\left(\tilde{F}_{\mu}, A\right)$ be a PVSES over a soft universe (U, Z). An *agree-possibility vague soft expert set* (agree-PVSES) over U, denoted as $\left(\tilde{F}_{\mu}, A\right)_1$ is a possibility vague soft expert subset of $\left(\tilde{F}_{\mu}, A\right)$ which is defined as:

$$\left(\tilde{F}_{\mu}, A\right)_1 = \left(\tilde{F}(\alpha), \mu(\alpha)\right), \qquad \text{where } \alpha \in E \times X \times \{1\}.$$

Definition 3.8. Let $\left(\tilde{F}_{\mu}, A\right)$ be a PVSES over a soft universe (U, Z). A *disagree-possibility vague soft expert set* (disagree-PVSES) over U, denoted as $\left(\tilde{F}_{\mu}, A\right)_0$ is a possibility vague soft expert subset of $\left(\tilde{F}_{\mu}, A\right)$ which is defined as:

$$\left(\tilde{F}_{\mu}, A\right)_0 = \left(\tilde{F}(\alpha), \mu(\alpha)\right), \qquad \text{where } \alpha \in E \times X \times \{0\}.$$

4   Basic Operations on Possibility Vague Soft Expert Sets

In this section, we introduce some basic operations on PVSES, namely the complement, AND, OR, union and intersection of PVSES and proceed to study some of the properties related to these operations.

Definition 4.1. Let $\left(\tilde{F}_{\mu}, A\right)$ be a PVSES over a soft universe (U, Z). Then the *complement* of $\left(\tilde{F}_{\mu}, A\right)$, denoted by $\left(\tilde{F}_{\mu}, A\right)^c$ is defined as:

$$\left(\tilde{F}_{\mu}, A\right)^c = \left(\tilde{c}\left(\tilde{F}(\alpha)\right), c(\mu(\alpha))\right), \qquad \forall a \in A,$$

where $\tilde{c}$ is a vague complement and c is a fuzzy complement.

Proposition 4.2. Let $\left(\tilde{F}_{\mu}, A\right)$ be a PVSES over a soft universe (U, Z). Then the following property holds true:

$$\left(\left(\tilde{F}_{\mu}, A\right)^c\right)^c = \left(\tilde{F}_{\mu}, A\right).$$

Proof. Suppose that $\left(\tilde{F}_{\mu}, A\right)$ is a PVSES over a soft universe (U, Z) defined as $\left(\tilde{F}_{\mu}, A\right) = \left(\tilde{F}(e), \mu(e)\right)$. Now let $\left(\tilde{F}_{\mu}, A\right)^c = \left(\tilde{G}_{\delta}, B\right)$. Then by Definition 4.1, $\left(\tilde{G}_{\delta}, B\right) = \left(\tilde{G}(e), \delta(e)\right)$ such that $\tilde{G}(e) = \tilde{c}\left(\tilde{F}(e)\right)$ and $\delta(e) = c(\mu(e))$. Thus it follows that $\left(\tilde{G}_{\delta}, B\right)^c = \left(\tilde{c}\left(\tilde{G}(e)\right), c(\delta(e))\right) = \left(\tilde{c}\left(\tilde{c}\left(\tilde{F}(e)\right)\right), c\left(c(\mu(e))\right)\right) = \left(\tilde{F}(e), \mu(e)\right) = \left(\tilde{F}_{\mu}, A\right)$. Therefore $\left(\left(\tilde{F}_{\mu}, A\right)^c\right)^c = \left(\tilde{G}_{\delta}, B\right)^c = \left(\tilde{F}_{\mu}, A\right)$. Hence it is proven that $\left(\left(\tilde{F}_{\mu}, A\right)^c\right)^c = \left(\tilde{F}_{\mu}, A\right)$. ∎

Definition 4.3. Let $\left(\tilde{F}_{\mu}, A\right)$ and $\left(\tilde{G}_{\delta}, B\right)$ be PVSESs over a soft universe (U, Z). Then the *union* of $\left(\tilde{F}_{\mu}, A\right)$ and $\left(\tilde{G}_{\delta}, B\right)$, denoted by $\left(\tilde{F}_{\mu}, A\right) \tilde{\cup} \left(\tilde{G}_{\delta}, B\right)$ is a PVSES defined as $\left(\tilde{F}_{\mu}, A\right) \tilde{\cup} \left(\tilde{G}_{\delta}, B\right) = \left(\tilde{H}_{\lambda}, C\right)$, where $C = A \cup B$ and

$$\lambda(\alpha) = \max\bigl(\mu(\alpha), \delta(\alpha)\bigr), \quad \forall\ \alpha \in C,$$

and
$$\tilde{H}(\alpha) = \tilde{F}(\alpha)\ \tilde{\cup}\ \tilde{G}(\alpha), \quad \forall\ \alpha \in C$$

where
$$\tilde{H}(\alpha) = \begin{cases} \tilde{F}(\alpha) & \alpha \in A - B, \\ \tilde{G}(\alpha) & \alpha \in B - A, \\ \max\left(\tilde{F}(\alpha), \tilde{G}(\alpha)\right) & \alpha \in A \cap B. \end{cases}$$

Proposition 4.4. *Let* $\left(\tilde{F}_\mu, A\right), \left(\tilde{G}_\delta, B\right)$ *and* $\left(\tilde{H}_\lambda, C\right)$ *be any three PVSES over a soft universe* (U, Z). *Then the following results hold true:*

(i) $\left(\tilde{F}_\mu, A\right) \tilde{\cup} \left(\tilde{G}_\delta, B\right) = \left(\tilde{G}_\delta, B\right) \tilde{\cup} \left(\tilde{F}_\mu, A\right)$

(ii) $\left(\tilde{F}_\mu, A\right) \tilde{\cup} \left(\left(\tilde{G}_\delta, B\right) \tilde{\cup} \left(\tilde{H}_\lambda, C\right)\right) = \left(\left(\tilde{F}_\mu, A\right) \tilde{\cup} \left(\tilde{G}_\delta, B\right)\right) \tilde{\cup} \left(\tilde{H}_\lambda, C\right)$

(iii) $\left(\tilde{F}_\mu, A\right) \tilde{\cup} \left(\tilde{F}_\mu, A\right) \subseteq \left(\tilde{F}_\mu, A\right)$

(iv) $\left(\tilde{F}_\mu, A\right) \tilde{\cup} \left(\tilde{\varnothing}_\mu, A\right) = \left(\tilde{F}_\mu, A\right)$

Proof. The proofs are straightforward.   ∎

Definition 4.5. Let $\left(\tilde{F}_\mu, A\right)$ and $\left(\tilde{G}_\delta, B\right)$ be PVSESs over a soft universe (U, Z). Then the *intersection* of $\left(\tilde{F}_\mu, A\right)$ and $\left(\tilde{G}_\delta, B\right)$, denoted by $\left(\tilde{F}_\mu, A\right) \tilde{\cap} \left(\tilde{G}_\delta, B\right)$ is a PVSES defined as $\left(\tilde{F}_\mu, A\right) \tilde{\cap} \left(\tilde{G}_\delta, B\right) = \left(\tilde{H}_\lambda, C\right)$, where $C = A \cup B$ and

$$\lambda(\alpha) = \min\bigl(\mu(\alpha), \delta(\alpha)\bigr), \quad \forall\ \alpha \in C,$$

and
$$\tilde{H}(\alpha) = \tilde{F}(\alpha)\ \tilde{\cap}\ \tilde{G}(\alpha), \quad \forall\ \alpha \in C$$

where
$$\tilde{H}(\alpha) = \begin{cases} \tilde{F}(\alpha) & \alpha \in A - B, \\ \tilde{G}(\alpha) & \alpha \in B - A, \\ \min\left(\tilde{F}(\alpha), \tilde{G}(\alpha)\right) & \alpha \in A \cap B. \end{cases}$$

Proposition 4.6. *Let* $\left(\tilde{F}_\mu, A\right), \left(\tilde{G}_\delta, B\right)$ *and* $\left(\tilde{H}_\lambda, C\right)$ *be any three PVSES over a soft universe* (U, Z). *Then the following results hold true:*

(i) $\left(\tilde{F}_\mu, A\right) \tilde{\cap} \left(\tilde{G}_\delta, B\right) = \left(\tilde{G}_\delta, B\right) \tilde{\cap} \left(\tilde{F}_\mu, A\right)$

(ii) $\left(\tilde{F}_\mu, A\right) \tilde{\cap} \left(\left(\tilde{G}_\delta, B\right) \tilde{\cap} \left(\tilde{H}_\lambda, C\right)\right) = \left(\left(\tilde{F}_\mu, A\right) \tilde{\cap} \left(\tilde{G}_\delta, B\right)\right) \tilde{\cap} \left(\tilde{H}_\lambda, C\right)$

(iii) $\left(\tilde{F}_\mu, A\right) \tilde{\cap} \left(\tilde{F}_\mu, A\right) \subseteq \left(\tilde{F}_\mu, A\right)$

(iv) $\left(\tilde{F}_\mu, A\right) \tilde{\cap} \left(\tilde{\varnothing}_\mu, A\right) = \left(\tilde{\varnothing}_\mu, A\right)$

Proof. The proofs are similar to that of Proposition 4.4 and are therefore omitted.   ∎

Proposition 4.7. *Let* $\left(\tilde{F}_\mu, A\right), \left(\tilde{G}_\delta, B\right)$ *and* $\left(\tilde{H}_\lambda, C\right)$ *be any three PVSES over a soft universe* (U, Z). *Then the following results hold true:*

(i) $\left(\tilde{F}_\mu, A\right) \tilde{\cup} \left(\left(\tilde{G}_\delta, B\right) \tilde{\cap} \left(\tilde{H}_\lambda, C\right)\right) =$
$\left(\left(\tilde{F}_\mu, A\right) \tilde{\cup} \left(\tilde{G}_\delta, B\right)\right) \tilde{\cap} \left(\left(\tilde{F}_\mu, A\right) \tilde{\cup} \left(\tilde{H}_\lambda, C\right)\right).$

(ii) $\left(\tilde{F}_{\mu}, A\right) \tilde{\cap} \left(\left(\tilde{G}_{\delta}, B\right) \tilde{\cup} \left(\tilde{H}_{\lambda}, C\right)\right) =$
$\left(\left(\tilde{F}_{\mu}, A\right) \tilde{\cap} \left(\tilde{G}_{\delta}, B\right)\right) \tilde{\cup} \left(\left(\tilde{F}_{\mu}, A\right) \tilde{\cap} \left(\tilde{H}_{\lambda}, C\right)\right).$

Proof. The proofs are straightforward by Definitions 4.3 and 4.5.   ∎

Proposition 4.8. *Let* $\left(\tilde{F}_{\mu}, A\right)$ *and* $\left(\tilde{G}_{\delta}, B\right)$ *be any two PVSES over a soft universe* (U, Z). *Then the De Morgan's laws hold true:*

(i) $\left(\left(\tilde{F}_{\mu}, A\right) \tilde{\cup} \left(\tilde{G}_{\delta}, B\right)\right)^{c} = \left(\tilde{F}_{\mu}, A\right)^{c} \tilde{\cap} \left(\tilde{G}_{\delta}, B\right)^{c}.$

(ii) $\left(\left(\tilde{F}_{\mu}, A\right) \tilde{\cap} \left(\tilde{G}_{\delta}, B\right)\right)^{c} = \left(\tilde{F}_{\mu}, A\right)^{c} \tilde{\cup} \left(\tilde{G}_{\delta}, B\right)^{c}.$

Proof. The proofs are straightforward by Definitions 4.1, 4.3 and 4.5.   ∎

Definition 4.9. Let $\left(\tilde{F}_{\mu}, A\right)$ and $\left(\tilde{G}_{\delta}, B\right)$ be PVSESs over a soft universe (U, Z). Then "$\left(\tilde{F}_{\mu}, A\right)$ *AND* $\left(\tilde{G}_{\delta}, B\right)$", denoted by $\left(\tilde{F}_{\mu}, A\right) \tilde{\wedge} \left(\tilde{G}_{\delta}, B\right)$ is a PVSES defined by

$$\left(\tilde{F}_{\mu}, A\right) \tilde{\wedge} \left(\tilde{G}_{\delta}, B\right) = \left(\tilde{H}_{\lambda}, A \times B\right),$$

where $\left(\tilde{H}_{\lambda}, A \times B\right) = \left(\tilde{H}(\alpha, \beta), \lambda(\alpha, \beta)\right)$, such that $\tilde{H}(\alpha, \beta) = \tilde{F}(\alpha) \cap \tilde{G}(\beta)$ and $\lambda(\alpha, \beta) = \min\left(\mu(\alpha), \delta(\beta)\right)$, for all $(\alpha, \beta) \in A \times B$ and $\cap$ represents the basic intersection.

Definition 4.10. Let $\left(\tilde{F}_{\mu}, A\right)$ and $\left(\tilde{G}_{\delta}, B\right)$ be PVSESs over a soft universe (U, Z). Then "$\left(\tilde{F}_{\mu}, A\right)$ *OR* $\left(\tilde{G}_{\delta}, B\right)$", denoted by $\left(\tilde{F}_{\mu}, A\right) \tilde{\vee} \left(\tilde{G}_{\delta}, B\right)$ is a PVSES defined by

$$\left(\tilde{F}_{\mu}, A\right) \tilde{\vee} \left(\tilde{G}_{\delta}, B\right) = \left(\tilde{H}_{\lambda}, A \times B\right),$$

where $\left(\tilde{H}_{\lambda}, A \times B\right) = \left(\tilde{H}(\alpha, \beta), \lambda(\alpha, \beta)\right)$, such that $\tilde{H}(\alpha, \beta) = \tilde{F}(\alpha) \cup \tilde{G}(\beta)$ and $\lambda(\alpha, \beta) = \max\left(\mu(\alpha), \delta(\beta)\right)$, for all $(\alpha, \beta) \in A \times B$ and $\cup$ represents the basic union.

Proposition 4.11. *Let* $\left(\tilde{F}_{\mu}, A\right), \left(\tilde{G}_{\delta}, B\right)$ *and* $\left(\tilde{H}_{\lambda}, C\right)$ *be any three PVSES over a soft universe* (U, Z). *Then the following properties hold true:*

(i) $\left(\tilde{F}_{\mu}, A\right) \tilde{\wedge} \left(\left(\tilde{G}_{\delta}, B\right) \tilde{\wedge} \left(\tilde{H}_{\lambda}, C\right)\right) = \left(\left(\tilde{F}_{\mu}, A\right) \tilde{\wedge} \left(\tilde{G}_{\delta}, B\right)\right) \tilde{\wedge} \left(\tilde{H}_{\lambda}, C\right).$

(ii) $\left(\tilde{F}_{\mu}, A\right) \tilde{\vee} \left(\left(\tilde{G}_{\delta}, B\right) \tilde{\vee} \left(\tilde{H}_{\lambda}, C\right)\right) = \left(\left(\tilde{F}_{\mu}, A\right) \tilde{\vee} \left(\tilde{G}_{\delta}, B\right)\right) \tilde{\vee} \left(\tilde{H}_{\lambda}, C\right).$

(iii) $\left(\tilde{F}_{\mu}, A\right) \tilde{\vee} \left(\left(\tilde{G}_{\delta}, B\right) \tilde{\wedge} \left(\tilde{H}_{\lambda}, C\right)\right) =$
$\left(\left(\tilde{F}_{\mu}, A\right) \tilde{\vee} \left(\tilde{G}_{\delta}, B\right)\right) \tilde{\wedge} \left(\left(\tilde{F}_{\mu}, A\right) \tilde{\vee} \left(\tilde{H}_{\lambda}, C\right)\right).$

(iv) $\left(\tilde{F}_{\mu}, A\right) \tilde{\wedge} \left(\left(\tilde{G}_{\delta}, B\right) \tilde{\vee} \left(\tilde{H}_{\lambda}, C\right)\right) =$
$\left(\left(\tilde{F}_{\mu}, A\right) \tilde{\wedge} \left(\tilde{G}_{\delta}, B\right)\right) \tilde{\vee} \left(\left(\tilde{F}_{\mu}, A\right) \tilde{\wedge} \left(\tilde{H}_{\lambda}, C\right)\right).$

Proof. The proofs are straightforward by Definitions 4.9 and 4.10.   ∎

Proposition 4.12. *Let* $\left(\tilde{F}_\mu, A\right)$ *and* $\left(\tilde{G}_\delta, B\right)$ *be any two PVSES over a soft universe* (U, Z). *Then the De Morgan's laws hold true:*

(i) $\qquad \left(\left(\tilde{F}_\mu, A\right) \tilde{\wedge} \left(\tilde{G}_\delta, B\right)\right)^c = \left(\tilde{F}_\mu, A\right)^c \tilde{\vee} \left(\tilde{G}_\delta, B\right)^c$

(ii) $\qquad \left(\left(\tilde{F}_\mu, A\right) \tilde{\vee} \left(\tilde{G}_\delta, B\right)\right)^c = \left(\tilde{F}_\mu, A\right)^c \tilde{\wedge} \left(\tilde{G}_\delta, B\right)^c$

Proof. The proofs are similar to that of Proposition 4.8 and are thus omitted.

5   Application of PVSESs in a Decision Making Problem

In this section, we present an algorithm to solve problems involving PVSESs.

Algorithm

1. Input the PVSES $\left(\tilde{F}_\mu, Z\right)$.
2. Find the values of $t_{\tilde{F}(z_i)}(u_i) - f_{\tilde{F}(z_i)}(u_i)$ for each element $u_i \in U$, where $t_{\tilde{F}(z_i)}(u_i)$ and $f_{\tilde{F}(z_i)}(u_i)$ are the truth membership function and false membership function of each of the elements $u_i \in U$ respectively.
3. Find the highest numerical grade for the agree-PVSES and disagree-PVSES.
4. Compute the score of each element $u_i \in U$ by taking the sum of the products of the numerical grade of each element with the corresponding degree of possibility μ_i, for the agree-PVSES and disagree-PVSES, denoted by A_i and D_i respectively.
5. Find the values of the score $r_i = A_i - D_i$ for each element $u_i \in U$.
6. Determine the value of the highest score, $s = \max_{u_i \in U} \{r_i\}$. Then the decision is to choose element u_i as the optimal or best solution to the problem. If there are more than one element with the highest r_i score, then any one of those elements can be chosen as the optimal solution.

Suppose that school A is looking to select the recipient of the Valedictorian award from the graduating batch of students for a particular year. The three shortlisted candidates from the graduating batch of students form the universe of elements, $U = \{u_1, u_2, u_3\}$. The selection committee consists of three school board members represented by the set $X = \{p, q, r\}$ (a set of experts) and the set $Q = \{1 = agree, 0 = disagree\}$ is the set of opinions of the selection committee members. The selection committee considers a set of parameters, $E = \{e_1, e_2, e_3, e_4\}$, where the parameters e_i $(i = 1, 2, 3, 4)$ represent the characteristics or qualities that the candidates are assessed on, namely "academic excellence", "active in extra-curricular activities", "good behaviour" and "highly disciplined" respectively. After interviewing all the candidates and going through their certificates and other supporting documents, the selection committee constructs the PVSES $\left(\tilde{F}_\mu, Z\right)$ which is given in matrix form in Table 1. Next the PVSES $\left(\tilde{F}_\mu, Z\right)$ is applied to the algorithm given above and used by the selection committee to determine the best student to be given the Valedictorian award for a particular year. The final scores of A_i and D_i which represents the score of each numerical grade for the agree-PVSES and disagree-PVSES respectively are given in Table 2. It was found that $s = \max_{u_i \in U} \{r_i\} = r_1$. Therefore, the selection committee should select student u_1 as the recipient of the Valedictorian award for that year. The other tables of values have been omitted due to space constraints.

$$\left(\tilde{F}_\mu, Z\right) = \left\{(e_1, p, 1) = \left\{\left(\frac{u_1}{[0.4, 0.6]}, 0.4\right), \left(\frac{u_2}{[0.5, 0.5]}, 0.3\right), \left(\frac{u_3}{[0.9, 1]}, 0.7\right)\right\}\right\}, ...,$$

Table 1. The PVSES $\left(\tilde{F}_\mu, Z\right)$

Agree	u_1	u_2	u_3	Disagree	u_1	u_2	u_3
$(e_1, p, 1)$	[0.4, 0.6], 0.4	[0.5, 0.5], 0.3	[0.9, 1], 0.7	$(e_2, p, 0)$	[0, 0.12], 0.75	[0.3, 0.7], 0.2	[0.8, 0.9], 0.6
$(e_2, p, 1)$	[0, 0.05], 0.9	[0.3, 0.4], 0.8	[0.2, 0.45], 0.15	$(e_4, p, 0)$	[0.8, 1], 0.62	[1, 1], 0.25	[0.3, 0.35], 0.9
$(e_4, p, 1)$	[0.4, 0.7], 0.1	[0.5, 0.75], 0.8	[0, 0.3], 0.3	$(e_1, q, 0)$	[0.3, 0.5], 0.55	[0.7, 0.9], 0.1	[1, 1], 0.99
$(e_1, q, 1)$	[0.8, 0.9], 0.3	[0.4, 0.5], 0.2	[0.6, 0.8], 0.3	$(e_3, q, 0)$	[0.2, 0.3], 0.2	[0.1, 0.4], 0.3	[0.5, 0.6], 0.1
$(e_2, q, 1)$	[0.6, 0.7], 0.6	[0.1, 0.1], 0.8	[0.9, 0.95], 0.5	$(e_4, q, 0)$	[0, 0.1], 0.6	[0.3, 0.6], 0.8	[0, 0.2], 0.1
$(e_3, q, 1)$	[0.9, 1], 0.1	[0.5, 0.6], 0.4	[0, 0.15], 0.5	$(e_1, r, 0)$	[0.3, 0.3], 0.3	[0.5, 0.7], 0.5	[0.4, 0.9], 0.3
$(e_4, q, 1)$	[0, 0.5], 0.3	[0.2, 0.5], 0.2	[0, 0.2], 0.9	$(e_2, r, 0)$	[0.2, 0.35], 0.9	[0.25, 0.7], 0.7	[0.1, 0.15], 0.3
$(e_1, r, 1)$	[0.3, 0.5], 0.7	[0.1, 0.4], 0.2	[0, 0], 1	$(e_3, r, 0)$	[0.2, 0.5], 0.1	[0.3, 0.7], 0.1	[0.1, 0.4], 0.6
$(e_3, r, 1)$	[0.15, 0.35], 0.6	[0.6, 0.9], 0.9	[0.4, 0.7], 0.1				
$(e_4, r, 1)$	[0.9, 0.95], 0.99	[0.2, 0.25], 0.8	[0, 0.1], 0.4				

Table 2. The score $r_i = A_i - D_i$

A_i	D_i	r_i
Score $(u_1) = 1.0015$	Score $u_1 = 0$	1.0015
Score $(u_2) = 0.35$	Score $u_2 = 0.135$	0.215
Score $(u_3) = 1.055$	Score $u_3 = 1.6$	-0.545

The theory of PVSES introduced here is a generalization of soft sets. Table 2 shows the result of the application of the algorithm to the PVSES model introduced here. The result provides us with not only an upper and lower bound of the membership function for an element but also a degree of possibility as well as a set of opinions given by a set of experts for each element in the universal set without the need for any additional operations. These special features sets the PVSES model apart from all other similar models in the literature and represents the advantages of the PVSES model.

6   Conclusion

In this paper the concept of possibility vague soft expert set which is a combination of the notion of possibility vague soft sets and soft expert sets was established. The basic operations and some of the fundamental properties of these sets are proved. Finally, an algorithm is introduced and the application of the PVSES model in a decision making problem is presented.

Acknowledgments. The author would like to gratefully acknowledge the financial assistance received from the Ministry of Education, Malaysia and UCSI University, Malaysia under Grant no. FRGS/1/2014/ST06/UCSI/03/1.

References

1. Molodtsov, D.: Soft Set Theory - First Results. Computers and Mathematics with Appl. 37, 19–31 (1999)
2. Maji, P.K., Biswas, R., Roy, A.R.: Fuzzy Soft Sets. Journal of Fuzzy Mathematics 3(9), 589–602 (2001)
3. Majumdar, P., Samanta, S.K.: Generalized Fuzzy Soft Sets. Computers and Mathematics with Applications 59(4), 1425–1432 (2010)
4. Alkhazaleh, S., Salleh, A.R., Hassan, N.: Soft Multiset Theory. Applied Mathematical Sciences 5(72), 3561–3573 (2011)
5. Alkhazaleh, S., Salleh, A.R.: Fuzzy Soft Multiset Theory. Abstract and Applied Analysis 2012, Article ID: 350603, 20 (2012)
6. Alkhazaleh, S., Salleh, A.R., Hassan, N.: Fuzzy Parameterized Interval-valued Fuzzy Soft Set. Applied Mathematical Sciences 5(67), 3335–3346 (2011)
7. Alkhazaleh, S., Salleh, A.R., Hassan, N.: Possibility Fuzzy Soft Sets. Advances in Decision Sciences 2011, Article ID: 479756, 18 (2011)
8. Alkhazaleh, S., Salleh, A.R.: Generalized Interval-valued Fuzzy Soft Sets. Journal of Applied Mathematics 2012, Article ID: 870504, 18 (2012)
9. Alkhazaleh, S., Salleh, A.R.: Soft Expert Sets. Advances in Decision Sciences 2011, Article ID: 757868, 12 (2011)
10. Alkhazaleh, S., Salleh, A.R.: Fuzzy Soft Expert Set and its Application. Applied Mathematics 5, 1349–1368 (2014)
11. Alhazaymeh, K., Hassan, N.: Vague Soft Multiset Theory. International Journal of Pure and Applied Mathematics 93(4), 511–523 (2014)
12. Alhazaymeh, K., Hassan, N.: Interval-Valued Vague Soft Sets and its Application. Advances in Fuzzy Systems 2012, Article ID: 208489, 7 (2012)
13. Alhazaymeh, K., Hassan, N.: Generalized Vague Soft Sets and its Application. International Journal of Pure and Applied Mathematics 77(3), 391–401 (2012)

14. Alhazaymeh, K., Hassan, N.: Possibility Vague Soft Set and its Application in Decision Making. International Journal of Pure and Applied Mathematics 77(4), 549–563 (2012)
15. Hassan, N., Alhazaymeh, K.: Vague soft expert set theory. In: AIP Conference Proceedings, vol. 1522, pp. 953–958 (2013)
16. Xu, W., Ma, J., Wang, S., Hao, G.: Vague Soft Sets and their Properties. Computers and Mathematics with Applications 59, 787–794 (2010)
17. Gau, W.L., Buehrer, D.J.: Vague Sets. IEEE Transactions on Systems, Man and Cybernetics 23(2), 610–614 (1993)

An Iterative Method for Solving Fuzzy Fractional Differential Equations

Ali Ahmadian[1], Fudziah Ismail[1], Norazak Senu[1], Soheil Salahshour[2],
Mohamed Suleiman[3], and Sarkhosh Seddighi Chaharborj[1]

[1] Department of Mathematics, Faculty of Science, Universiti Putra Malaysia,
43400 UPM, Serdang, Selangor, Malaysia
[2] Young Researchers and Elite Club, Mobarakeh Branch, Islamic Azad University,
Mobarakeh,Iran.
[3] Institute for Mathematical Research (INSPEM), Universiti Putra Malaysia,
43400 UPM, Serdang, Selangor, Malaysia
ahmadian.hosseini@gmail.com,
soheilsalahshour@yahoo.com,
{fudziah,norazak,mohamed,sarkhosh}@upm.edu.my

Abstract. The aim of this paper is to solve fuzzy fractional differential equations (FFDEs) of the Caputo type. The basic idea is to convert FFDEs to a type of fuzzy Volterra integral equation. Then the obtained Volterra integral equation will be exploited with some suitable quadrature rules to get a fractional predictor-corrector method. The results show that the proposed method exhibit high precision with low cost.

Keywords: Fuzzy fractional differential equations, Fuzzy Caputo differentiability, Fuzzy Volterra integral equation, Predictor-Corrector method.

1 Introduction

Fractional differential equations (FDEs) is an old topic of mathematics because the idea for this subject was planted over 300 years ago, but during the last few decades it has found many interesting applications in solving real-world problems possessing the power law effect [1, 2]. Due to the accuracy of fractional calculus in modeling various engineering and physical phenomena [3–5]. During last decade, researchers devoted much efforts on the various numerical simulation techniques for the solution of FDEs [6–12].

In the recent years, following the Agarwal et al.'s paper [13] which was presented the conception of solutions for fractional differential equations with uncertainty, the theoretical and numerical aspects of fuzzy fractional differential equations (FFDEs) have been studied by some authors such [14–19] and [20–27].

In the numerical treatment of ordinary differential equations , Adams methods represent one of the most used and studied class of implicit (Adams-Moulton) and explicit (Adams-Bashforth) linear multistep methods. The wide popularity of Adams methods is mainly due to their good stability properties, reasonable computational cost and ease of implementation. For this reason, several efforts

© Springer Science+Business Media Singapore 2015
M.W. Berry et al. (Eds.): SCDS 2015, CCIS 545, pp. 88–96, 2015.
DOI: 10.1007/978-981-287-936-3_9

have been dedicated to generalize Adams methods to FDEs (e.g. see [12, 28]); indeed the presence of a persistent memory, and the consequent increase of the computational effort needed for evaluating the solution away from the origin, requires the development of efficient algorithms. As it was stated earlier, a few numerical methods for FFDEs have been presented in the literature. Therefore, the main aim of this paper is to develop an easy implemented fractional predictor-corrector (FPC) method , discussed in [12], for FFDEs and investigate the error analysis of the method for solving this type of FDEs.

This paper is constructed as follows: In Section 2, the basic notations of the fuzzy sets and fuzzy Caputo fractional derivative for fuzzy functions are recalled. In Section 3, we develop the FPC solution method for FFDEs under the Caputo generalized Hukuhara differentiability. A test problem will be solved in Section 4 by FPC method to demonstrate the accuracy and validity. Finally, some conclusions are drawn.

2   Basic Concepts

In this part, some preliminaries related to fuzzy fractional differential equations are provided. For more details see [4, 14, 15, 29, 30].

Let $\mathbb{R}$ be a set of real number. We recall that a fuzzy number represents a mapping $\omega : \mathbb{R} \to [0, 1]$ fulfilling following properties:

(a) ω is upper semi-continuous,
(b) ω is fuzzy convex, i.e., $\omega(\lambda x + (1 - \lambda)y) \geq min\{\omega(x), \omega(y)\}$ for all $x, y \in \mathbb{R}, \lambda \in [0, 1]$,
(c) ω is normal, i.e.,$\exists x_0 \in \mathbb{R}$ for which $\omega(x_0) = 1$,
(d) supp $\omega = \{x \in \mathbb{R} \mid \omega(x) > 0\}$ is the support of the ω, and its closure cl(supp ω) is compact.

Let $\mathcal{F}$ be the set of all fuzzy number on $\mathbb{R}$. The r-level set of a fuzzy number $u \in \mathcal{F}$, $0 \leq r \leq 1$, denoted by $[\omega]_r$, is defined as

$$[\omega]_r = \begin{cases} \{x \in \mathbb{R} \mid \omega(x) \geq r\} & if \quad 0 < r \leq 1 \\ cl(supp\ \omega) & if \quad r = 0 \end{cases}$$

We notice that the r-level set of a fuzzy number is a closed and bounded interval $[\underline{\omega}(r), \overline{\omega}(r)]$, where $\underline{\omega}(r)$ is the left-hand endpoint of $[\omega]_r$ and $\overline{\omega}(r)$ represents the right-hand endpoint of $[\omega]_r$.

A corresponding definition using a parametric form was introduced in [30] as:

Definition 1. *A fuzzy number ω in parametric form is a pair $(\underline{\omega}, \overline{\omega})$ of functions $\underline{\omega}(r)$, $\overline{\omega}(r)$, $0 \leq r \leq 1$, which satisfy the following requirements:*

1. *$\underline{\omega}(r)$ is a bounded non-decreasing left continuous function in $(0, 1]$, and right continuous at 0,*
2. *$\overline{\omega}(r)$ is a bounded non-increasing left continuous function in $(0, 1]$, and right continuous at 0,*
3. *$\underline{\omega}(r) \leq \overline{\omega}(r)$, $0 \leq r \leq 1$.*

The Hausdorff distance between fuzzy numbers is given by $d : \mathcal{F} \times \mathcal{F} \longrightarrow [0, \infty]$,

$$\mathcal{H}(\omega, \gamma) = \sup_{r \in [0,1]} \max\{|\underline{\omega}(r) - \underline{\gamma}(r)|, |\overline{\omega}(r) - \overline{\gamma}(r)|\},$$

where $\omega = (\underline{\omega}(r), \overline{\omega}(r))$, $v = (\underline{\gamma}(r), \overline{\gamma}(r)) \subset \mathbb{R}$ is utilized in [29]. Then, it is easy to see that $\mathcal{H}$ is a metric in $\mathcal{F}$ and has the following properties (see for example [31])

(i) $\mathcal{H}(\omega + \mu, \gamma + \mu) = \mathcal{H}(\omega, \gamma)$, $\quad \forall \omega, \gamma, \mu \in \mathcal{F}$,
(ii) $\mathcal{H}(k\omega, k\gamma) = |k|\mathcal{H}(\omega, \gamma)$, $\quad \forall\, k \in \mathbb{R}, \omega, \gamma \in \mathcal{F}$,
(iii) $\mathcal{H}(\omega + \gamma, \mu + \zeta) \leq \mathcal{H}(\omega, \mu) + \mathcal{H}(\gamma, \zeta)$, $\quad \forall \omega, \gamma, \mu, \zeta \in \mathcal{F}$,
(iv) $(\mathcal{H}, \mathcal{F})$ is a complete metric space.

The H-derivative (differentiability in the sense of Hukuhara)for fuzzy-set-valued functions was at first established by Puri and Ralescu in [31] and it is based on the H-difference of fuzzy sets, as follows.

Definition 2. *Let $x, y \in \mathcal{F}$. If there exists $z \in \mathcal{F}$ such that $\omega = \mu + \gamma$, then γ is called the H-difference of ω and μ, and it is denoted by $\omega \ominus \mu$.*

It is worth noting that the sign "$\ominus$" always arises for H-difference, and also, $\omega \ominus \mu \neq \omega + (-1)\mu$.

Let us recall the definition of strongly generalized differentiability introduced in [32].

Definition 3. *Let $\xi : (a, b) \to \mathcal{F}$ and $x_0 \in (a, b)$. We say that ξ is strongly generalized differential at x_0, if there exists an element $\xi'(\tau_0) \in \mathcal{F}$, such that*

(i) for all $h > 0$ sufficiently small, $\exists \xi(\tau_0 + h) \ominus \xi(\tau_0)$, $\exists \xi(\tau_0) \ominus h\xi(\tau_0 - h)$ and the limits (in the metric $\mathcal{H}$)

$$\lim_{h \searrow 0} \frac{\xi(\tau_0 + h) \ominus \xi(\tau_0)}{h} = \lim_{h \searrow 0} \frac{\xi(\tau_0) \ominus \xi(\tau_0 - h)}{h} = \xi'(\tau_0)$$

or

(ii) for all $h > 0$ sufficiently small, $\exists \xi(\tau_0) \ominus \xi(\tau_0 + h)$, $\exists \xi(\tau_0 - h) \ominus \xi(\tau_0)$ and the limits (in the metric $\mathcal{H}$)

$$\lim_{h \searrow 0} \frac{\xi(\tau_0) \ominus \xi(\tau_0 + h)}{-h} = \lim_{h \searrow 0} \frac{\xi(\tau_0 - h) \ominus \xi(\tau_0)}{-h} = \xi'(\tau_0)$$

or

(iii) for all $h > 0$ sufficiently small, $\exists \xi(\tau_0 + h) \ominus \xi(\tau_0)$, $\exists \xi(\tau_0 - h) \ominus \xi(\tau_0)$ and the limits (in the metric $\mathcal{H}$)

$$\lim_{h \searrow 0} \frac{\xi(\tau_0 + h) \ominus \xi(\tau_0)}{h} = \lim_{h \searrow 0} \frac{\xi(\tau_0 - h) \ominus \xi(\tau_0)}{-h} = \xi'(\tau_0)$$

or

(iv) for all $h > 0$ sufficiently small, $\exists \xi(\tau_0) \ominus \xi(\tau_0 + h)$, $\exists \xi(\tau_0) \ominus \xi(\tau_0 - h)$ and the limits (in the metric $\mathcal{H}$)

$$\lim_{h \searrow 0} \frac{\xi(\tau_0) \ominus \xi(\tau_0 + h)}{-h} = \lim_{h \searrow 0} \frac{\xi(\tau_0) \ominus \xi(\tau_0 - h)}{h} = \xi'(\tau_0)$$

Before proceed, we express $\mathcal{C}([a,b], \mathcal{F})$ as the space of all continuous fuzzy-valued functions on $[a,b]$. Also, $\mathcal{L}([a,b], \mathcal{F})$ stands for the space of all Lebesque integrable fuzzy-valued functions on the bounded interval $[a,b] \subset \mathbb{R}$.

Now, the Riemann-Liouville integral of fuzzy-valued function is determined as follows:

Definition 4. *(see,[14]) Let $\xi \in \mathcal{C}([a,b], \mathcal{F}) \cap \mathcal{L}([a,b], \mathcal{F})$. The fuzzy Riemann-Liouville integral of fuzzy-valued function f is defined as follows:*

$$\left(I_{a+}^{\beta} \xi\right)(x) = \frac{1}{\Gamma(\beta)} \int_a^x \frac{\xi(t)dt}{(x-t)^{1-\beta}}, \quad x > a, \ 0 < \beta \leq 1. \tag{1}$$

Let us assume the r-cut depiction of fuzzy-valued function ξ as $\xi(x,r) = [\underline{\xi}(x,r), \overline{\xi}(x,r)]$, for $0 \leq r \leq 1$, then it is indicated that the Riemann-Liouville integral of fuzzy-valued function ξ is based on the lower and upper functions as follows:

Theorem 1 *(see,[14]). Let $\xi \in \mathcal{C}([a,b], \mathcal{F}) \cap \mathcal{L}([a,b], \mathcal{F})$ is a fuzzy-valued function. The Riemann-Liouville integral of a fuzzy-valued function ξ be disclosed as follows:*

$$\left(I_{a+}^{\beta} \xi\right)(x;r) = \left[\left(I_{a+}^{\beta} \underline{\xi}\right)(x;r), \left(I_{a+}^{\beta} \overline{\xi}\right)(x;r)\right], \quad 0 \leq r \leq 1, \tag{2}$$

where

$$\left(I_{a+}^{\beta} \underline{\xi}\right)(x;r) = \frac{1}{\Gamma(\beta)} \int_a^x \frac{\underline{\xi}(t;r)dt}{(x-t)^{1-\beta}},$$

$$\left(I_{a+}^{\beta} \overline{\xi}\right)(x;r) = \frac{1}{\Gamma(\beta)} \int_a^x \frac{\overline{\xi}(t;r)dt}{(x-t)^{1-\beta}}. \tag{3}$$

Now, the fuzzy Caputo fractional derivatives about order $0 < \beta \leq 1$ for fuzzy-valued function ξ was defined by Salahshour et al. [14].

Definition 5. *(see, [14]). Let $\xi \in \mathcal{C}([a,b], \mathcal{F}) \cap \mathcal{L}([a,b], \mathcal{F})$ be a fuzzy set-value function, then ξ is a Caputo fuzzy H-differentiable at x when:*

$$(^C D_{a+}^{\beta} \xi)(x) = \frac{1}{\Gamma(1-\beta)} \int_a^x \frac{\xi'(t)}{(x-t)^{\beta}} dt, \tag{4}$$

where $0 < \beta \leq 1$; then, we say f is $^C[(1) - \beta]$-differentiable if Eq. (4) holds while ξ is (1)-differentiable, and f is $^C[(2) - \beta]$-differentiable if Eq. (4) holds while f is (2)-differentiable.

3   Solution Method

In this section we shall develop the fundamental algorithm that we intend to apply for the solution of fuzzy initial value problems with fuzzy Caputo derivatives. The algorithm is a generalization of the fractional Adams-Bashforth-Moulton integrator that is well known for the numerical solution of FDEs [12]. For this purpose, let us consider the fuzzy linear fractional relaxation-oscillation problem as:

$$\begin{cases} {}_{0}^{c}D_{t}^{\alpha}X(t) = -BX(t) + g(t) \simeq F(t, X(t)), \\ \quad\quad\quad X(0) = X_0 \in E, \end{cases} \tag{5}$$

where X_0 is a fuzzy initial vector condition, B is relaxation coefficient and $\alpha \in (0, 1]$.

It is easy to verify that this problem is equivalent to the following fuzzy Volterra integral equation under ${}^{C}[(1) - \alpha]$-differentiability

$$X(t) = X_0 + \frac{1}{\Gamma(\alpha)} \int_0^t (t - \xi)^{\alpha-1}[-BX(\xi) + g(\xi)]d\xi,$$

and under ${}^{C}[(2) - \alpha]$-differentiability, we have

$$X(t) = X_0 \ominus (-1)\frac{1}{\Gamma(\alpha)} \int_0^t (t - \xi)^{\alpha-1}[-BX(\xi) + g(\xi)]d\xi.$$

We have exploited a fractional Adams-Bashforth as predictor and a fractional Adams-Moulton as corrector formulas based on the algorithm presented in [12]. Here, we state the numerical method under ${}^{C}[(2) - \beta]$-differentiability. Analogously to the demonstration of the technique under ${}^{C}[(2) - \beta]$-differentiability, one can extend it for ${}^{C}[(1) - \beta]$-differentiability.

The predictor value $X_h^P(t_{n+1})$ is obtained by the fuzzy fractional Adams-Bashforth method as follows:

$$X_h^P(t_{n+1}) = X_0 \ominus (-1)\frac{1}{\Gamma(\alpha)} \sum_{j=0}^{n} b_{j,n+1} f(t_j, X_h(t_j)), \tag{6}$$

where

$$b_{j,n+1} = \frac{h^\alpha}{\alpha}[(n + 1 - j)^\alpha - (n - j)^\alpha]. \tag{7}$$

Also, the following equation then gives us our corrector formula (i.e., the fuzzy fractional variant of the one-step Adams-Moulton method), which is

$$X_h(t_{n+1}) = X_0 \ominus (-1)\frac{h^\alpha}{\Gamma(\alpha+2)} \left(\sum_{j=0}^{n} a_{j,n+1} f(t_j, X_h(t_j)) + [f(t_j, X_h^P(t_j))] \right), \tag{8}$$

where

$$
a_{j,n+1} = \begin{cases}
n^{\alpha-1} - (n-\alpha)(n+1)^{\alpha}, & j = 0, \\
(n-j+2)^{\alpha+1} + (n-j)^{\alpha+1} - 2(n-j+1)^{\alpha+1}, & 1 \le j \le n, \\
1, & j = n+1.
\end{cases}
\tag{9}
$$

It is worth noting that we have used a uniform discrete scheme $t_j = jh$, $j = 0, 1, ..., n$ and $T = nh$, where T is the final time.

Our algorithm, the fuzzy fractional AdamsBashforthMoulton method, is fully described now by Equations (6) and (8) with the weights $a_{j,n+1}$ and $b_{j,n+1}$ being defined according to (7) and (9), respectively.

4   Numerical Experiment

In this section the numerical solution of linear FFDEs is illustrated by means of the FPC method presented in Section 4.

Consider the following fuzzy time-fractional Bloch equation:

$$
\begin{cases}
{}_0^C D_t^\alpha X(t) = -BX(t), \\
X(0) = [0.5 + 0.5r, 1.5 - 0.5r]
\end{cases}
\tag{10}
$$

where $0 < \alpha \le 1$, $B = 1$. The exact solution under $^C[(2) - \beta]$-differentiability is obtained as follows:

$$
\begin{cases}
\underline{X}(x;r) = (0.5 + 0.5r)E_{\alpha,1}[-x^\alpha], & 0 < \alpha \le 1, \\
\overline{X}(x;r) = (1.5 - 0.5r)E_{\alpha,1}[-x^\alpha], & 0 < r \le 1,
\end{cases}
\tag{11}
$$

in which $E_{\alpha,1}$ is the Mittag-Leffler function and $B = 1$.

Now, in order to obtain the numerical solution based on the FPC method in compare of the given exact solution, we assume different values of α and derive the absolute errors at $T = 1$ graphed in Fig. 1, i.e.

$$
[N_e]^r = [N_{1e}(1;r), N_{2e}(1;r)] = \left[|\underline{X}_n(1;r) - \underline{X}(1;r)|, |\overline{X}_n(1;r) - \overline{X}(1;r)| \right].
$$

Also, the absolute errors of the Eq. (10) with the assumption $\alpha = 0.95$ is depicetd in Fig. (2). From the figures, one can easily conclude that the method can achieve a high accuracy for any order of $0 < \alpha \le 1$. Moreover, as α approaches 1, the error decreases gradually and is in high agreement with the exact solution.

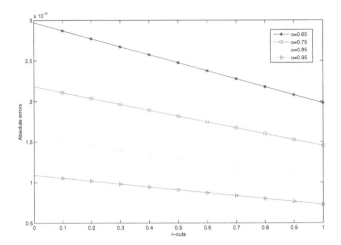

Fig. 1. $N_{2e}(1;r)$ for different α, with $h = 0.01$ and $T = 1$

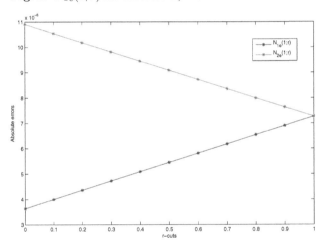

Fig. 2. $[N_e]^r$ for $\alpha = 0.95$, with $h = 0.01$ and $T = 1$

5   Conclusion

In this work, a high accuracy FPC method is used to solve FFDEs. The fractional derivative is considered in the fuzzy Caputo sense. The numerical results obtained by the proposed technique are compared with exact solution to illustrate validity and applicability of the proposed technique. From the numerical results, it is obvious that the proposed method exhibit high precision and easy-implemented. The numerical results was obtained by Matlab R2011a.

References

1. Oldham, K.B., Spainer, J.: The Fractional Calculus: Theory and Applications of Differentiation and Integration to Arbitrary Order. Academic Press, New York (1974)
2. Kilbas, A.A., Srivastava, H.M., Trujillo, J.J.: Theory and Applications of Fractional Differential Equations. Elsevier, Amsterdam (2006)
3. Hilfer, R.: Application of Fractional Calculus in Physics. World Scientific, Singapore (2000)
4. Baleanu, D., Diethelm, K., Scalas, E., Trujillo, J.J.: Fractional Calculus Models and Numerical Methods. World Scientific Publishing Company (2012)
5. Golmankhaneh, A.K., Yengejeh, A.M., Baleanu, D.: On the fractional Hamilton and Lagrange mechanics. International Journal of Theoretical Physics 51, 2909–2916 (2012)
6. Lim, S.C., Eab, C.H., Mak, K.H., Li, M., Chen, S.Y.: Solving linear coupled fractional differential equations by direct operational method and some applications. Mathematical Problems in Engineering, 1–28 (2012)
7. Pedas, A., Tamme, E.: On the convergence of spline collocation methods for solving fractional differential equations. Journal of Computational and Applied Mathematics 235, 3502–3514 (2011)
8. Jiang, W.H.: Solvability for a coupled system of fractional differential equations at resonance. Nonlinear Analysis: Real World Applications 13, 2285–2292 (2012)
9. Inc., M.: The approximate and exact solutions of the space-and time-fractional Burger's equations with initial conditions by VIM. J. Math. Anal. Appl. 345, 476–484 (2008)
10. Sweilam, N.H., Khader, M.M.: Exact solutions of some coupled nonlinear partial differential equations using the homotopy perturbation method. Comput. Math. Appl. 58, 2134–2141 (2009)
11. Ding, X.L., Jiang, Y.L.: Waveform relaxation methods for fractional differential-algebraic equations with the Caputo derivatives. Fract. Calc. Appl. Anal. 17, 585–604 (2014)
12. Diethelm, K., Ford, N.J., Freed, A.D.: A predictor-corrector approach for the numerical solution of fractional differential equations. Nonlinear Dynamics 29, 3–22 (2002)
13. Agarwal, R.P., Lakshmikantham, V., Nieto, J.J.: On the concept of solution for fractional differential equations with uncertainty. Nonlinear Anal. 72, 2859–2862 (2010)
14. Salahshour, S., Allahviranloo, T., Abbasbandy, S., Baleanu, D.: Existence and uniqueness results for fractional differential equations with uncertainty. Advances in Difference Equations 2012, 112 (2012)
15. Salahshour, S., Allahviranloo, T., Abbasbandy, S.: Solving fuzzy fractional differential equations by fuzzy Laplace transforms, Commun. Nonlinear. Sci. Numer. Simulat: 17, 1372–1381 (2012)
16. Salahshour, S., Ahmadian, A., Senu, N., Baleanu, D., Agarwal, P.: On Analytical Solutions of the Fractional Differential Equation with Uncertainty: Application to the Basset Problem. Entropy 17, 885–902 (2015)
17. Ahmadian, A., Suleiman, M., Salahshour, S., Baleanu, D.: A Jacobi operational matrix for solving fuzzy linear fractional differential equation. Adv. Difference Equ. 2013, 104 (2013)

18. Ahmadian, A., Suleiman, M.: An Operational Matrix Based on Legendre Polynomials for Solving Fuzzy Fractional-Order Differential Equations. Abstract and Applied Analysis 2013, Article ID 505903, 29 (2013)
19. Balooch Shahriyar, M.R., Ismail, F., Aghabeigi, S., Ahmadian, A., Salahshour, S.: An Eigenvalue-Eigenvector Method for Solving a System of Fractional Differential Equations with Uncertainty. Mathematical Problems in Engineering 2013, Article ID 579761, 11 (2013)
20. Ghaemi, F., Yunus, R., Ahmadian, A., Salahshour, S., Suleiman, M., Faridah Saleh, S.: Application of Fuzzy Fractional Kinetic Equations to Modelling of the Acid Hydrolysis Reaction. Abstract and Applied Analysis 2013, Article ID 610314, 19 (2013)
21. Ahmadian, A., Senu, N., Larki, F., Salahshour, S., Suleiman, M., Islam, S.: Numerical solution of fuzzy fractional pharmacokinetics model arising from drug assimilation into the blood stream. Abstract and Applied Analysis 2013, Article ID 304739 (2013)
22. Ahmadian, A., Senu, N., Larki, F., Salahshour, S., Suleiman, M., Shabiul Islam, M.: A Legendre Approximation for Solving a Fuzzy Fractional Drug Transduction Model into the Bloodstream. In: Herawan, T., Ghazali, R., Deris, M.M. (eds.) Recent Advances on Soft Computing and Data Mining SCDM 2014. AISC, vol. 287, pp. 25–34. Springer, Heidelberg (2014)
23. Ahmadian, A., Salahshour, S., Baleanu, D., Amirkhani, H., Yunus, R.: Tau method for the numerical solution of a fuzzy fractional kinetic model and its application to the Oil Palm Frond as a promising source of xylose. Journal of Computational Physics 294, 562–584 (2015)
24. Mazandarani, M., Vahidian Kamyad, A.: Modified fractional Euler method for solving Fuzzy Fractional Initial Value Problem, Commun. Nonlinear Sci. Numer. Simulat. 18, 12–21 (2013)
25. Alikhani, R., Bahrami, F.: Global solutions for nonlinear fuzzy fractional integral and integro-differential equations. Nonlinear. Sci. Numer. Simulat. 18, 2007–2017 (2013)
26. Malinowski, M.T.: Random fuzzy fractional integral equations-theoretical foundations. Fuzzy Sets Syst. 265, 39–62 (2015)
27. Ngo, V.H.: Fuzzy fractional functional integral and differential equations. Fuzzy Sets Syst. doi:10.1016/j.fss.2015.01.009 (In press)
28. Garrappa, R.: On linear stability of predictor-corrector algorithms for fractional differential equations. Int. J. Comput. Math. 87, 2281–2290 (2010)
29. Wu, H.C.: The improper fuzzy Riemann integral and its numerical integration. Information Science 111, 109–137 (1999)
30. Dubios, D., Prade, H.: Towards fuzzy differential calculus-part3. Fuzzy Sets and Systems 8, 225–234 (1982)
31. Puri, M.L., Ralescu, D.: Differential for fuzzy function. Journal of Mathematical Analysis and Applications 91, 552–558 (1983)
32. Bede, B., Gal, S.G.: Generalizations of the differentiability of fuzzy-number-valued functions with applications to fuzzy differential equations. Fuzzy Sets Syst. 151, 581–599 (2005)

Contrast Comparison of Flat Electroencephalography Image: Classical, Fuzzy, and Intuitionistic Fuzzy Set

Suzelawati Zenian[1], Tahir Ahmad[2,*], and Amidora Idris[1]

[1] Department of Mathematical Sciences, Faculty of Science,
Universiti Teknologi Malaysia, 81310 UTM, Johor Bahru, Malaysia
`suzelawati@gmail.com`, `amidora@utm.my`
[2] Centre for Sustainable Nanomaterials, Ibnu Sina Institute for Scientific
and Industrial Research, Universiti Teknologi Malaysia, 81310 UTM, Johor Bahru, Malaysia
`tahir@ibnusina.utm.my`

Abstract. Image processing is used to enhance visual appearance of images for further interpretation. One of the applications of image processing is in medical imaging. Generally, the pixel values of an image may not be precise as uncertainty arises within the gray values of an image due to several factors. In this paper, the image of Flat EEG (fEEG) is compared via classical, fuzzy, and intuitionistic fuzzy set (IFS) methods. Furthermore, the comparison between the input and output images of fEEG is carried out based on contrast comparison.

Keywords: Flat EEG, intuitionistic fuzzy set, fuzzy set, hesitation degree, fuzzy image.

1   Introduction

The revolution in medical imaging gives high impact in the development towards modern medical care. Various kinds of methods have been implemented in medical imaging that involve classical and fuzzy approaches. The importance to implement fuzzy approach is that medical image itself contains a lot of uncertainties. Much effort has been carried out to enhance ambiguous medical images. Therefore, it is relevant to implement alternative tool such as fuzzy set in order to describe, analyze, and interpret image.

Fuzzy set which is a generalization of classical set, was introduced by Zadeh in 1965. It is a powerful tool in dealing with uncertainties and provide a formal way of describing real-world phenomena. The extension of fuzzy set was introduced by Atanassov in 1983 known as intuitionistic fuzzy set (IFS) [1]. The advantage of IFS compared to fuzzy set is that it considers more uncertainties in terms of membership and non-membership functions. Intuitionistic fuzzy image processing consists of five main steps which are fuzzification, intuitionistic fuzzification, modification of intuitionistic fuzzy components, intuitionistic defuzzification, and defuzzification [2].

In the process of imaging and transformation such as fEEG, it is hard to avoid the inheritance of different kinds of noise during recording of the EEG signals. Uncertainties may arise within every transformation since the regions of clusters in fEEG are

© Springer Science+Business Media Singapore 2015
M.W. Berry et al. (Eds.): SCDS 2015, CCIS 545, pp. 97–105, 2015.
DOI: 10.1007/978-981-287-936-3_10

not always defined. The main aim of this paper is to improve the visibility of the clusters of epileptic foci by using contrast enhancement together with IFS approach. The electrical potential that occurs as the vague boundary of the epileptic foci in the input image will be reduced. The results show that IFS method is able to obtain clearer boundary of the epileptic foci compared to the input image.

2   Basic Concepts

Fuzzy Topographic Topological Mapping (FTTM) which is introduced by [3] is a novel non-invasive technique for solving neuromagnetic inverse problem. It is based on mathematical concepts namely topology and fuzzy. It aims to accommodate static simulated, experimental magnetoencephalography (MEG), and recorded electroencephalography (EEG) signals. FTTM consists of three algorithms that link four components of the model as shown in Fig. 1. The four components namely the Magnetic Contour Plane (MC), Base Magnetic Plane (BM), Fuzzy Magnetic Field (FM), and Topographic Magnetic Field (TM).

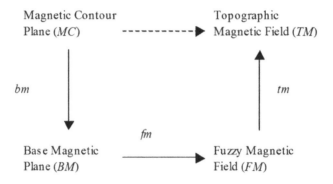

Fig. 1. FTTM [3].

EEG is a system that measures and records the electrical activity of the brain in graphic form [4]. It reads voltage differences on the head relative to a given point. Fig. 2 shows a sample of EEG signal during seizure.

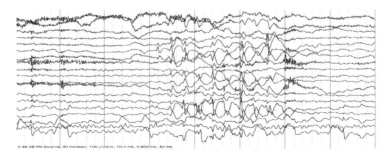

Fig. 2. EEG signal [5].

In order to map the high dimensional signal, namely EEG, into low dimensional space, a method which is known as fEEG has been developed by the Fuzzy Research Group of UTM in 1999 [5]. This method has been used purely for visualization and able to preserve information recorded during seizure. The details of fEEG is given as follows:

Fauziah's EEG coordinate system [5] is defined as

$$C_{EEG} = \left\{ \left((x,y,z), e_p \right) : x,y,z,e_p \in \Re \text{ and } x^2 + y^2 + z^2 = r^2 \right\} \tag{1}$$

whereby r is the radius of a patient head. The mapping of C_{EEG} to a plane is defined as $S_t : C_{EEG} \to MC$ such that

$$S_t \left((x,y,z), e_p \right) = \left(\frac{rx + iry}{r+z}, e_p \right) = \left(\frac{rx}{r+z}, \frac{ry}{r+z} \right)_{e_p(x,y,z)} \tag{2}$$

where $MC = \left\{ \left((x,y)_0, e_p \right) : x,y,e_p \in R \right\}$ is the first component of FTTM. Both C_{EEG} and MC were designed and proven as 2-manifolds.

Meanwhile S_t is designed to be a one to one function as well as being conformal. Details of proofs are contained in [5]. The EEG signal during seizure can be compressed to Fig. 3 and analyzed second by second as Fig. 4. In Fig. 4, the position of cluster centers in a patient are in green colour. Meanwhile, the red colour represents the location of sensors on the surface of the patient's head.

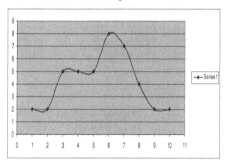

Fig. 3. Compressed EEG signal [5].

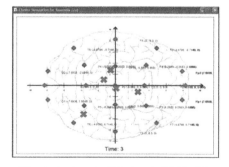

Fig. 4. Analyzed EEG signal (fEEG) [5].

Fauziah [5] transformed the EEG signal into fEEG via the flattening method. Furthermore, Abdy and Ahmad [6] transformed the fEEG into image by using fuzzy approach. There are three main steps that are involved in the transformation of fEEG into image [6].

a)   fEEG is divided into pixels (see Fig. 5)

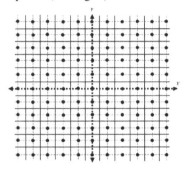

Fig. 5. fEEG pixels [6].

b)   The membership value for each pixel is determined in a cluster centre and the maximum operator of fuzzy set is implemented (see Fig. 6)

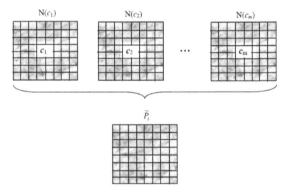

Fig. 6. Fuzzy neighborhood of each cluster centre c_j of a fEEG [6].

c)   The membership value of pixel is transformed into image data (see Fig. 7)

Fig. 7. fEEG image (input image) [6].

3   Methodology

Let A be an IFS in a finite set $X = \{x_1, x_2, \ldots, x_n\}$ which is defined as $A = \{(x, \mu_A(x), \nu_A(x)) | x \in X\}$ whereby $\mu_A(x), \nu_A(x): X \to [0,1]$ represent the membership and non-membership respectively. The necessary conditions that must be fulfilled are as follows [1]

$$0 \le \mu_A(x) + \nu_A(x) \le 1 \tag{3}$$

and

$$\pi_A(x) = 1 - \mu_A(x) - \nu_A(x) \tag{4}$$

such that $0 \le \pi_A(x) \le 1$.

In IFS, the sum of the membership and non-membership values will not always equal to one. Therefore, there exists parameter $\pi_A(x)$ that makes the summation equal to one. According to [1], the occurrence of $\pi_A(x)$ is due to the lack of knowledge about the membership degree or personal error while calculating the distances between two fuzzy sets. Thus the membership value will lie in the interval $\left[\mu_A(x) - \pi_A(x), \mu_A(x) + \pi_A(x)\right]$ because of the hesitation that occurs in the membership function.

In 2012, [7] proposed a method which is based on IFS known as the window based enhancement scheme (WBES). It is aimed to enhance the contrast of medical images. Therefore, in this paper, the window based enhancement scheme is applied in order to obtain an enhanced image of fEEG. However, there is a slight difference in the initial step with the proposed step by Chaira [7]. In [7], the image is initially divided into 4 partitioned windows and fuzzification is carried out for each partitioned window. According to [7], a lot of noise is presented on increasing the numbers of partitioned windows.

On the other hand, in this paper, the fEEG image is initially undergoing the fuzzification process which is applied to the entire image and later the image is divided into 4 partitioned windows. It is known as the revised WBES version for fEEG. This is to ensure that this method will result in better output fEEG image. The revised algorithm is described as follows:

Algorithm

1. *The entire input image is initially fuzzified by using*

$$\mu_A(g_{ij}) = \frac{g_{ij} - g_{\min}}{g_{\max} - g_{\min}} \tag{5}$$

2. *The image is divided into 4 partitioned windows and enhancement is carried out for each partitioned window.*

3. *The non-membership function is computed by using Sugeno type intuitionistic fuzzy generator as follows*

$$v_A\left(g_{ij}\right) = \frac{1 - \mu_A\left(g_{ij}\right)}{1 + \lambda \mu_A\left(g_{ij}\right)}, \quad \lambda > 0 \tag{6}$$

4. *The hesitation degree is obtained from (4)*

$$\pi_A\left(g_{ij}\right) = 1 - \mu_A\left(g_{ij}\right) - \frac{1 - \mu_A\left(g_{ij}\right)}{1 + \lambda \mu_A\left(g_{ij}\right)} \tag{7}$$

5. *The mean of each partitioned window is calculated*

6. *The modified membership value is given by*

$$\mu_A^{\;mod}\left(g_{ij}\right) = \mu_A\left(g_{ij}\right) - mean\ window \times \pi_A\left(g_{ij}\right) \tag{8}$$

7. *Finally, the contrast enhancement is applied to each partitioned window by using the intensifier operator as given by (9)*

$$\mu_A^{\;enh}\left(g_{ij}\right) = \begin{cases} 2\left[\mu_A^{\;mod}\left(g_{ij}\right)\right]^2 & if \quad \mu_A^{\;mod}\left(g_{ij}\right) \leq 0.5 \\ 1 - 2\left[1 - \mu_A^{\;mod}\left(g_{ij}\right)\right]^2 & if \quad 0.5 < \mu_A^{\;mod}\left(g_{ij}\right) \leq 1 \end{cases} \tag{9}$$

In the algorithm, g_{ij} is the $(i,j)^{th}$ gray level of the image.

4   Results

The aforementioned algorithm is implemented on fEEG image during epileptic seizure. Moreover, the results are compared with other different methods in classical, fuzzy, and IFS (WBES) by [7]. The classical methods in contrast enhancement are the histogram equalization (global) and the adaptive histogram equalization (local).

According to [8], histogram equalization is a technique whereby it changes the gray-level distribution of an image. In global histogram equalization, the computed transformation function is applied to all pixels of the input image. The output is a uniform resulting histogram with the same percentage of pixels in every gray level. Whereas the local histogram equalization will enhance details in small areas within an image.

The fuzzy method is based on [9] that the main aim is to obtain a contrast enhanced image. The IFS (WBES) by [7] is also being compared whereby the image is initially divided into 4 partitioned windows before implementing the fuzzification process. The image of fEEG that is used as the input image is at time 1 of size 201x201 (see Fig. 7). There are two clusters of electrical current sources whereby the brightness represents the strength of the electrical potential.

By implementing these methods on the input image, resulting in the output of fEEG image as shown in Fig. 8 till Fig. 10. By using the global histogram equalization, it seems that the lighter area spreads out wider and only one cluster can be observed. Local histogram equalization shows that the two clusters are quite blur and not in good quality.

Moreover, there is not much differences in the fEEG image by applying fuzzy and revised IFS (WBES) methods. But by implementing the IFS (WBES) method by [7], the output image shows complication of some noise. The revised IFS (WBES) shows a clearer viewed boundary (prominent) of the two cluster centres compared to the classical, fuzzy, and IFS (WBES) by [7]. The value of λ should be positive ($\lambda > 0$) for the IFS (WBES) and the revised version of IFS (WBES). In this paper, some values of λ will be used, namely when $\lambda = 0.01$, $\lambda = 1$, and $\lambda = 10$. According to [7], as λ increases the non-membership value will decrease and the hesitation degree will increase. Hence, the enhanced image starts to deteriorate as λ increases.

Furthermore, the measure of distinction between the input and output image is carried out in terms of contrast comparison. According to [10], the contrast comparison between a reference image x and a test image y is given as follows:

$$c(x, y) = \frac{2\sigma_x \sigma_y}{\sigma_x^2 + \sigma_y^2} \tag{10}$$

whereby σ_x and σ_y are the standard deviations of x and y. The standard deviations estimate the contrast between the images such that it measures how similar the contrast are. The range value is in the interval $[0,1]$ with the best value 1 if and only if the value of the standard deviations are the same. Table 1 shows the contrast comparison between the input and output image of fEEG via the compared methods.

a) Global Histogram Equalization   b) Local Histogram Equalization

Fig. 8. Classical method.

Fig. 9. Fuzzy method.

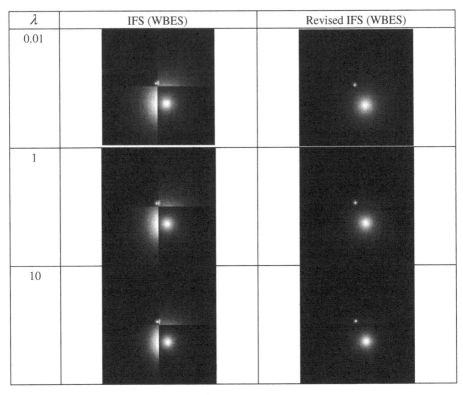

Fig. 10. IFS method.

Table 1. Contrast comparison of different methods.

Methods	Contrast comparison	
Global Histogram Equalization	0.5948	
Local Histogram Equalization	0.9788	
Fuzzy	0.9918	
IFS (WBES)	$\lambda = 0.01$	0.8757
	$\lambda = 1$	0.9075
	$\lambda = 10$	0.9436
Revised IFS (WBES)	$\lambda = 0.01$	0.9921
	$\lambda = 1$	0.9995
	$\lambda = 10$	0.9945

5   Conclusions

The fEEG image has been enhanced by using different methods ranging from the classical, fuzzy, and IFS. The resulting images are compared based on the contrast comparison. The IFS (WBES) does not give promising output to the fEEG image. A slight modification needs to be done in the initial step. Revised IFS (WBES) gives the highest value that is closed to 1 as $\lambda = 1$. The implementation of IFS helps to enhance the image of fEEG especially in visualizing the domain of electrical current sources. This is of outmost important to help neurologist to visualize and detect the epileptic foci clearly to overcome the problematic cells. The revised IFS (WBES) method is proven to be better than IFS (WBES) in enhancing the fEEG image.

Acknowledgements. The authors would like to thank their family members for their support and encouragement, the members of the Fuzzy Research Group (FRG), Department of Mathematics and Ibnu Sina Institute for Fundamental Science Studies, UTM for their assistance and cooperation. The first author wants to thank the Ministry of Education, Malaysia and Universiti Malaysia Sabah for granting permission to pursue her studies under SLAI.

References

1. Atanassov, K.T.: Intuitionistic Fuzzy Sets. J. Fuzzy Sets and Systems 20, 87–96 (1986)
2. Vlachos, I.K., Sergiadis, G.D.: Intuitionistic Fuzzy Image Processing. In: Nachtegael, M., Van der Weken, D., Kerre, E., Philips, W. (eds.) Soft Computing in Image Processing. STUDFUZZ, vol. 210, pp. 383–414. Springer, Heidelberg (2007)
3. Ahmad, T., Ahmad, R.S., Zakaria, F., Yun, L.L.: Development of Detection Model for Neuromagnetic Fields. Proceeding of BIOMED, pp. 119–121. Kuala Lumpur (2000)
4. Rudman, J.: EEG Technician. National Learning Corporation (2012)
5. Zakaria, F.: Dynamic Profiling of EEG Data during Seizure using Fuzzy Information. Ph.D thesis, Universiti Teknologi Malaysia (2008)
6. Abdy, M., Ahmad, T.: Transformation of EEG Signals into Image Form during Epileptic Seizure. Int. J. of Basic and Appl. Sc. 11, 18–23 (2011)
7. Chaira, T.: Medical Image Enhancement using Intuitionistic Fuzzy Set. In: IEEE 1st Int. Conf. on Recent Advances in Information Technology (RAIT 2012), pp. 54–57 (2012)
8. Marques, O.: Practical Image and Video Processing using MATLAB. John Wiley & Sons (2011)
9. Chaira, T., Ray, A.K.: Fuzzy Image Processing and Applications with MATLAB. CRC Press, Inc. (2009)
10. Wang, Z., Alan, C.B.: A Universal Image Quality Index. IEEE Signal Processing Letters 9(3), 81–84 (2002)

An Autocatalytic Model of a Pressurized Water Reactor in a Nuclear Power Generation

Azmirul Ashaari[1], Tahir Ahmad[2,*], Mustaffa Shamsuddin[2],
and Wan Munirah Wan Mohammad[1]

[1] Department of Mathematical Science, Faculty of Science,
UniversitiTeknologi Malaysia, 81310 UTM,Skudai, Johor, Malaysia
[2] Centre for Sustainable Nanomaterials, IbnuSina Institute for Scientific
and Industrial Research, UniversitiTeknologi Malaysia,
81310 UTM,Skudai, Johor, Malaysia
tahir@ibnusina.utm.my

Abstract. The control system of a nuclear reactor ensures the safe operation of a nuclear power plant. A Pressurized Water Reactor (PWR) in a nuclear reactor is a complex system since it contains uranium oxide. The aim of this paper is to model the process of operation for primary system in PWR in the form of graphical representation. The method of autocatalytic set approach of PWR has been introduced and presented in this paper. Further, the result of the dynamic process of the model has been presented, which is then is verified against published data.

Keywords: Pressurized Water Reactor (PWR), Graphical model, Autocatalytic Set.

1   Introduction

A pressurized water reactor (PWR) is one of the nuclear reactors that use light water as a coolant and moderator [1]. During the operation of PWR, the moderator remains in a liquid state despite the high temperature inside the reactor due to the high pressure in the primary coolant loop. PWR contains two parts namely primary and secondary systems as shown in Fig.1. However, the aim of this paper is to present the process of operation for primary system of PWR.

The PWR presented in [2] is used in this study. The operations for primary and secondary systems of PWR are described separately. This paper provides information on the primary system since the system is more complex due to the fission process. The flow of moderator begins in the reactor vessel where the moderator is being heated by nuclear fission. In general, the fission process occurs when atom U-235 captures the neutron. Subsequently atom U-235 separates into two major compounds as shown in Fig 2 [1].

Further, the hot moderator in the reactor vessel is being transferred to the pressurizer. The role of pressurizer is to control the temperature of moderator and the

© Springer Science+Business Media Singapore 2015
M.W. Berry et al. (Eds.): SCDS 2015, CCIS 545, pp. 106–115, 2015.
DOI: 10.1007/978-981-287-936-3_11

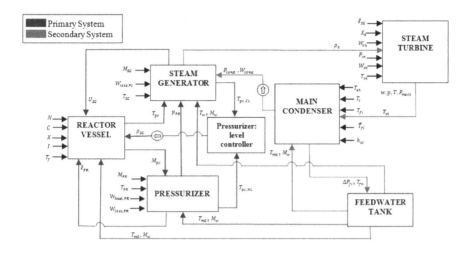

Fig. 1. Pressurizer Water Reactor Block Diagram System

pressure at certain level before the moderator is transferred to steam generator. The moderator in the secondary system absorbed the heat produced from the hot moderator in the primary system to form the phase change from water to steam. The moderator in primary system is then transferred back into the reactor vessel via reactor coolant pump. This process is repeated until the PWR reactor is shut down.

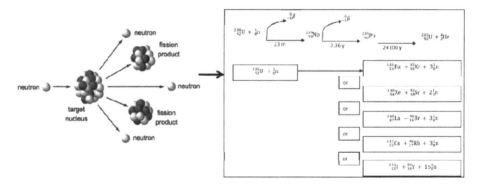

Fig. 2. Fission Product for U-235 and U-238

In order to understand and model the primary system, some instructions and methods have been noted from previous researchers. According to Babuska [3] under-standing the behavior and nature with a suitable mathematical approach is important in modelling the system. Meanwhile, Lindskog [4] stated that each system that has been modelled and designed should have contained every element that existed in the system. Hence, Tahir et. al. [5,6] presented mathematical modelling for each com-ponent of primary system in PWR. Further, according to Tahir et. al. [7] work, the concept of modelling of the system can also be presented in the form of a graph whereby each point of the graph represent the compounds that existed or were found in the system with the links that connect between each point. This method has been used to model the incineration process in Malaka and has proven successful in describing the behavior and operation of the model system. Therefore, the graph concept of an autocatalytic set (ACS) has been presented the next section.

2   Autocatalytic Set

Graph is a one of mathematics methods developed from the connection of linked points. The development of graph theory was first introduced by Swiss mathematician Leonhard Euler in order to solve the Seven Bridges of Königsberg problem [8]. The problem was to find a walkthrough of the city, where islands can only be reached via bridges. However, the walk path must not cross each bridge more than once, and every bridge must be completely crossed every time. According to Balakrishnan and Ranganathan [9] the development of the graph approached has grown into a significant area of mathematical researches and has also been used in other disciplines such as physics, chemistry, psychology, sociology and computer sciences. Further, Harary [10] described that graph theory has improved the mathematical technique where the interconnection between elements of natural and manmade can be modelled. The concept of graph is defined as the networks of points or nodes that are connected by links [9]. According to Epp [11] the characteristic of a graph can be described as a set of lines that are used to connect a set of points. The definition of directed graph was presented [8] as follows:

Definition 1.1. A directed graph $G = G(V, E)$ is defined by a set V of "vertices" and a set E of "edge" where each link is an ordered pair of vertices.

The set of vertices and edge can be represented as $V = \{v_1, v_2, v_3, ..., v_n\}$ and $E = \{e_1, e_2, e_3, ..., e_n\}$ with $C = (c_{ij})$ is the adjacency matrix of a graph. Nodes and links are also known as vertices and edge, respectively. The graph is defined as a connected graph if every pair of vertices in the graph is joined by edge.

Definition 1.2. Adjacency matrix of graph $G = G(V, E)$ with n vertices is a $n \times n$ matrix, whereby $c_{ij} = 1$ if E contains a directed link (i, j) and $c_{ij} = 0$ otherwise (for arrow from vertices j to i) for the entries of matrix $C = (c_{ij})$. The adjacency matrix of graph G can be described as follows:

$$C_{ij} \begin{cases} 1 & \text{if}(i,j) \in E \\ 0 & \text{if}(i,j) \notin E \end{cases} \tag{1}$$

where the entries matrix C_{ij} represent the connection between the jth and i^{th} vertices. Fig. 3 shows the example of graph C for four vertices with its adjacency matrix.

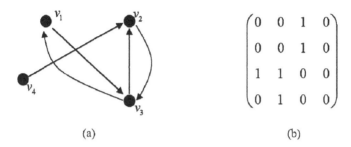

(a)   (b)

Fig. 3. (a) A graph for four vertices   (b) An adjacency matric of graph

Furthermore, the graph can also be defined as an Eulerian graph. The definition of Eulerian graph was stated in [8] as follows:

Definition 1.3. The graph is called an Euler circuit if the graph is connected and every vertex has even degree.

Autocatalytic set is defined as a set of catalytically react to the set of compounds where the member of the set catalyst each other (Eigen 1971; Kaufman 1971; Rossler 1971). Further, Jain and Krishna introduce the definition of an autocatalytic set in the form of the graph theoretical concept.

Definition 1.4. An autocatalytic set (ACS) is a sub graph, each of whose nodes have at least one incoming link from a node belonging to the same sub graph (see Fig. 4).

(a)   (b)

Fig. 4.   (a) A 1-cycle, the simplest ACS.   (b)   An irreducible graph but not cycle.

Further, according to Jain and Krishna [12] work, the concept of a dynamic system of a graph is defined as the couplings specified by the network of interaction from the set of coupled differential equation. The dynamic system for the set of coupling differential equation is presented as follows:

$$x_i' = \sum_{j=1}^{n} c_{ij} x_j - x_i \sum_{j,k=1}^{n} c_{kj} x_j \tag{2}$$

where contain vertices at fixed graph. The dynamic preserves the normalization of as stated below:

$$\sum_{i=1}^{n} x_i' = 0 \tag{3}$$

3   Modelling Graph Representation of a Primary System

The data for graph model of a primary system is taken from AP1000 nuclear power plant [13, 14]. The six vertices are fuel, moderator, corrosions, boron, nitrogen, and chlorides as shown in Table 1. Meanwhile, the existence and relationships between each vertices as presented in [1, 13, 15, and 16] is used as reference.

Table 1. The Vertices Parameter Graph

Vertices	Parameter	Vertices	Parameter
v_1	Fuel	v_4	Boron
v_2	Moderator (H_2O)	v_5	Nitrogen
v_3	Corrosion	v_6	Chlorides

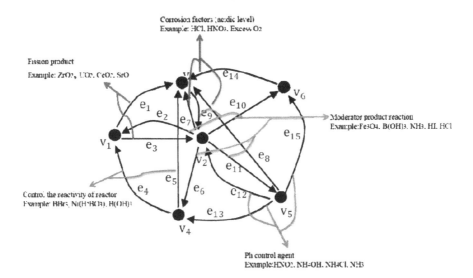

Fig. 5. An Autocatalytic Set of Primary System   (Gp)

R. Castelli [16] stated that the compounds of Co, Zn, Ni, Fe and Zr are classified as corrosion products. However, in this paper the term corrosion is used rather than corrosion products. The graph Gp shows that each link has the same color and thickness due to the same value of connectivity between the vertices in the graph as shown in Fig. 5.

Graph Gp shows the connection or links between each vertices. Each vertices play an important role in maintaining the operation of the primary system. Next section will discuss and describe more details about the dynamic process of graph representation of primary system or graph Gp.

4   Results and Discussions

The matrix for graph Gp is an irreducible adjacency matrix. Since Gp graph is an irreducible, Gp is an autocatalytic set [7]. However, Gp is not a Eulerian graph since every vertex has different in-degree and out-degree. Table 2 shows the degree of vertices for graphGp . The eigenvalues and eigenvectors for Gp are determined using Matlab.

Table 2. The Degree Vertices of Graph Gp

Vertices	V_1	V_2	V_3	V_4	V_5	V_6
In degree	2	3	5	2	1	2
Out degree	2	4	1	2	4	1

$$\text{Adjacency matrix of GraphGp} = \begin{pmatrix} 0 & 1 & 0 & 1 & 0 & 0 \\ 1 & 0 & 1 & 0 & 1 & 0 \\ 1 & 1 & 0 & 1 & 1 & 1 \\ 0 & 1 & 0 & 0 & 1 & 0 \\ 0 & 1 & 0 & 0 & 0 & 0 \\ 0 & 1 & 0 & 0 & 1 & 0 \end{pmatrix}$$

$$\text{Eigenvalue, } \lambda(C) = \begin{pmatrix} 2.4142 \\ -0.5000 + 0.8600i \\ -0.5000 - 0.8600i \\ -0.4142 \\ -1.0000 \\ 0.0000 \end{pmatrix}$$

$$\text{Eigenvector,} X = \begin{pmatrix} -0.3247 \\ -0.4943 \\ -0.6639 \\ -0.2895 \\ -0.2047 \\ -0.2895 \end{pmatrix}, \begin{pmatrix} -0.0000 + 0.0000i \\ 0.0000 - 0.3780i \\ 0.6547 \\ -0.0000 + 0.3780i \\ -0.3273 + 0.1890i \\ -0.0000 + 0.3780i \end{pmatrix}, \begin{pmatrix} -0.0000 - 0.0000i \\ 0.0000 + 0.3780i \\ 0.6547 \\ -0.0000 - 0.3780i \\ -0.3273 - 0.1890i \\ -0.0000 - 0.3780i \end{pmatrix},$$

$$\begin{pmatrix} 0.6113 \\ -0.0574 \\ -0.7260 \\ -0.1958 \\ 0.1385 \\ -0.1958 \end{pmatrix}, \begin{pmatrix} 0.5000 \\ -0.5000 \\ -0.5000 \\ 0.0000 \\ 0.5000 \\ 0.0000 \end{pmatrix}, \begin{pmatrix} 0.5774 \\ -0.0000 \\ -0.5774 \\ 0.0000 \\ -0.0000 \\ -0.5774 \end{pmatrix}$$

The eigenvalue which is larger than to all other eigenvalues is called PerronFrobenius [15] eigenvalue, namely

$$\lambda(C) = (2.4142)$$

Corresponding to $\lambda(C)$, the eigenvector consisting only of real and non-negative components will be known as PerronFrobenius eigenvector [17]. Here,

$$|X(C)| = \begin{pmatrix} 0.3247 \\ 0.4943 \\ 0.6639 \\ 0.2895 \\ 0.2047 \\ 0.2895 \end{pmatrix}$$

Further, according to Jain and Krishna [18, 19], if the subgraph of PFE is non-zero, then the adjacency matrix is represented as an entire graph. Hence, this situation is similar to the condition for adjacency matrix for graph Gp, which is then indicated as representing the whole process of graph Gp. The matrix A is the concentration percentage of compound for data of weight from AP1000 nuclear power plant [13, 14]. The percentage is calculated to ensure the sum is equal to 1. In order to get the rate of change for PWR, the matrix A and Gp are used. The rate of change for Gp is calculated using equation (2) and (3) and is shown in Table 3.

$$A = \begin{pmatrix} 0.6441 \\ 0.0960 \\ 0.1651 \\ 0.0031 \\ 0.0381 \\ 0.0537 \end{pmatrix}$$

Further, the negative sign for the fuel value shows that the compound is decreasing in volume due to the use of fuel in the process. Meanwhile, the rate of change for Gp   and PFE shows that the early exhaust variable is similar to nitrogen compounds and was the first to be depleted. The concept of selection of existing compounds in a primary system over time followed the technique proposed by Noor AinyHarith et al. [7].

During the operation in PWR, the nitrogen is depleted first. Nitrogen compounds deplete due to activation of oxygen in the coolant, resulting in the formation of nitrogen-16. Nitrogen-16 is a strong gama radiation emitter with a short half-life of 7.11 seconds. Next, chlorine is depleted due to the presence of nitrogen. Gilbert Gedeon et al. [20] described that if there is chlorine, the feed and bleed method should be performed to prevent corrosion. Further, boron compounds are depleted due to the use in controlling reactivity of fuel in primary system.

Table 3. The Dynamic System of Graph Gp

Element	The Rate of Change PWR	EigenVector	Descriptions
Fuel Moderator Corrosion Boron Nitrogen Chlorine	−1.2829 0.6414 0.4809 0.1275 0.0142 0.0189	0.3247 0.4943 0.6639 0.2895 0.2047 0.2895	Nitrogen depleted
Fuel Moderator Corrosion Boron Chlorine	Unspecified	0.3754 0.5164 0.6820 0.2522 0.2522	Chlorine depleted
Fuel Moderator Corrosion Boron	Unspecified	0.3216 0.5690 0.6851 0.3216	Boron depleted
Fuel Moderator Corrosion	Unspecified	0.4004 0.6479 0.6479	Fuel depleted
Moderator Corrosion	Unspecified	0.7071 0.7071	At the end of the process, moderator and corrosion exist after shut down of the primary system [15,16].

The data in [13] shows that boron levels were monitored and added at approximately 100 gallons per minute (22.71 m3/hour) into the moderator. Next fuel compound was stated as depleted, which indicates that the fuel compounds have been completely used in the operation of PWR. Finally, at the end of the operation, only the corrosion and moderator are left. The existing moderator is present in larger quantities than other compounds. Moreover, the need of a moderator to exist until the end is to ensure sufficient coolant in order to prevent PWR from overheating. On the other hand, corrosion exists due to the chemical reaction during the long term operation of PWR. The data from [13] shows that products due to corrosion are more abundant than the others when the operation of PWR was shut down. GraphGp   showed that the dynamic process has some similarity with the real-life process of operation in a primary system of PWR. Hence, graph Gp   has successfully presented the process and flow for the primary system of PWR.

5   Conclusion

A primary system of PWR is modeled by a graph. The necessary steps to model it as an autocatalytic set are shown in this paper. The simulated result of the model demonstrates that the dynamic processes of PWR are comparable to the published data.

Acknowledgment. This work has been supported by IbnuSina Institute, MyBrain15 from Ministry of High Education Malaysia and University Teknologi Malaysia.

References

1. Murray, R.: Nuclear Energy: An Introduction to the Concepts, Systems, and Applications of Nuclear Processes, Butterworth-Heinemann (2000)
2. Fazekas, C., Szederkenyi, G., Hangos, K.: A Simple Dynamic Model Of The Primary Circuit In VVER Plants For Controller Design Purposes. Nuclear Engineering and Design 237, 1071–1087 (2007)
3. Babuska, R.: Fuzzy Modelling: Principles, Methods and Application. In: Bonivento, C., Fantuzzi, C., Rovatti, R. (eds.) Fuzzy Logic Control-Advance in Methodology. World Scientific, Singapore (1998)
4. Lindskog, P.: Fuzzy Identification from a Grey Box Modeling Point of View. In: Hellendoorn, H., Driankov, D. (eds.) Fuzzy Model Identification, pp. 3–50. Springer (1997)
5. Ashaari, A., Ahmad, T., Shamsuddin, M., Adib Abdullah, M.: State space modeling of reactor core in a pressurized water reactor. In: Proceedings of the 21st National Symposium on Mathematical Sciences (SKSM21): Germination of Mathematical Sciences Education and Research towards Global Sustainability, vol. 1605, pp. 494–499. AIP Publishing (2014)
6. Ashaari, A., Ahmad, T., Shamsuddin, M., Omar, N.: Modelling Steam Generator System of Pressurized Water Reactor Using Fuzzy State Space. International Journal of Pure and Applied Mathematics 103(1) (2015)
7. Ahmad, T., Baharun, S., Arshad, K.A.: Modeling A Clinical Incineration Process using Fuzzy Autocatalytic Set. Journal of Mathematical Chemistry 47, 1263–1273 (2010)
8. Carlson, S.C.: Graph Theory. Encyclopedia Britannica (2010)
9. Balakrishnan, R., Ranganathan, K.A.: Textbook of Graph Theory. Springer (2012)
10. Harary, F.: Graph Theory. Addison Wesley Publishing Company, California (1969)
11. Epp, S.S.: Discrete Mathematics with Applications. PWS Publishing Company, Boston (1993)

12. Jain, S., Krishna, S.: Formation and Destruction of Autocatalytic Sets in an Evolving Network Model. Indian Institute of Science-Bangalore, India, Ph.D. Thesis (2003b)
13. EPS-GW-GL-700 Rev 1.: AP1000 European Design Control Document. Westinghouse Electric Company LLC, Pittsburgh (2009)
14. UKP-GW-GL-058.: UK AP1000 D1 Form Submission. Pittsburgh: Westinghouse Electric Company LLC (2009)
15. Neeb, K.H.: The Radiochemistry Of Nuclear Power Plants With Light Water Reactors. Walter de Gruyter (1997)
16. Castelli, R.: Nuclear Corrosion Modeling: The Nature of CRUD, Butterworth-Heinemann (2009)
17. Abu Bakar, S., Ahmad, T., Baharun, S.: Transition Probability Matrix for Fuzzy Autocatalytic Set of Fuzzy Graph Type-3. Advances and Applications in Mathematical Sciences 5, 25–38 (2010)
18. Jain, S., Krishna, S.: Autocatalytic Sets and the Growth of Complexity in an Evolutionary Model. Physical Review Letters 81, 5684–5687 (1998)
19. Jain, S., Krishna, S.: Emergence and Growth of Complex Networks in Adaptive Systems. Computer Physics Communications 121–122, 116–121 (1999)
20. Gilbert Gedeon, P.: Fundamentals of Chemistry. Nuclear engineering and technology, Continuing Education and Development, Inc. (1993)

Part III
Evolutionary Computing / Optimization

Selfish Gene Image Segmentation Algorithm

Noor Elaiza Abd Khalid, Norharyati Md Ariff, Ahmad Firdaus Ahmad Fadzil,
and Noorhayati Mohamed Noor

Faculty of Computer and Mathematical Sciences
Universiti Teknologi MARA Shah Alam Selangor, Malaysia
{elaiza,noorhayati}@tmsk.uitm.edu.my, sakurayati@yahoo.com,
firdausfadzil@melaka.uitm.edu.my

Abstract. The research proposes a selfish gene image segmentation algorithm as an alternative to Genetic Algorithm. Research in Genetic Algorithms originated from Darwin's theory faced the problem of finding the optimal solution due to its inherent characteristic of genetic drift and premature convergence. Selfish gene views genes as the basic unit in evolution. Thus the color image segmentation algorithm is designed based on virtual population with collection of genes rather than fixed genes chromosomes. The genes are positioned into predetermined loci forming two chromosomes that make up the virtual population in each generation. The chromosomes are rewarded and penalized according to the chromosomes performance. Evaluation with the ground truth images shows that the selfish gene is able to detect the variation of colors very similar to the way eye detect color.

Keywords: Selfish Gene Algorithm, Selfish Gene Theory, Richard Dawkins, Evolutionary Algorithm.

1   Introduction

Evolutionary algorithms (EA) are a family of population-based stochastic search and optimization methods inspired from the Darwinian theory [1]. Generally, these algorithms include creating an initial population of feasible solutions which evolves iteratively from generation to generation based on natural selection towards optimal solutions [3].

Currently the most well known and popular EAs, are genetic algorithm (GA) [10,11,18]. GA involves balancing the exploitation (finding better solution) and exploration (investigating unknown and new terrain) in the search space [12]. GA has been successfully applied in image segmentation problems [13,14].

In 1976, Richard Dawkins presented a new theory of evolution named as Selfish gene. His theory suggests genes as the basic unit of evolution as opposed to individual chromosomes [15] and the selection mechanism is express as gene-centred evolution [1]. The population is depicted as pools of genes and fight for the opportunity to survive as genotype in a chromosome [3,4] pursued this idea to proposed a new evolutionary optimization strategy called Selfish gene algorithm (SFGA). In contrast with

© Springer Science+Business Media Singapore 2015
M.W. Berry et al. (Eds.): SCDS 2015, CCIS 545, pp. 119–128, 2015.
DOI: 10.1007/978-981-287-936-3_12

genetic algorithm, the individual chromosomes are only virtually created during the process of fitness evaluation, crossover and mutation. Instead the genes are kept in a separate gene population pool named as population pool [2,19].

A comparison shows that SFGA performs better that GA in four algorithmic features. The first problem in GA is that the genes in an individual chromosomes are fixed. This may eliminate potential good genes in early generation and limits the problem space to non-optimal region [21]. Parameter tuning and balancing in the alleles in a fixed individual chromosomes to obtain optimum solutions are intricate [22]) as chromosomes improved throughout each generation but not the genes [24]. This phenomenon reduces the diversity of the population throughout each generation 17] and causes premature convergence and not yield the optimum solutions[23].

SFGA on the other hand, focuses on gene not the individual/chromosome. The good genes will be identified in fitness evaluation through the penalize and reward process. Thus, SFGA have a smaller number of parameters to tune.

In successive iterations of the algorithm, fitness-based selection takes place within the population of solutions. Better solutions are preferentially selected for survival into the next generation of solutions, with a diversity being introduced to the selected solutions in an attempt to uncover even better solutions over the next generation, with an aim of searching for a global optimum.

In the Selfish Gene Theory, which is very useful ideas from Richard Dawkins [3], the population can be simply seen as a pool of genes and the individual's genes fight for their appearances in the genotype of the vehicles. The survival of the fittest is a battle fought by genes, not the individual. In previous works, SFG theory was migrated into the field of evolutionary computation and produces a new algorithm called Selfish Gene Algorithms [3,4,5]. The researcher found that this new algorithm needs to be explored more because it gives very positive result on the test given. Therefore, this algorithm was early applied in the field of electronic computer aided design (CAD), for determining the logic for a BIST (Built-In Self-Test) architecture based on Cellular Automata [4,6,7,8,9]. It has also been introduced as data clustering by utilizing the segregation biological concept of distorted genes which can be recognized as a different species [20].

The organization of the rest of this paper is as follows: Section 2 presents our methods, including the overview of the methodology and descriptions of selfish gene algorithm structure. Section 3 discusses our results and discussions. Finally, we present our conclusion in Section 4.

2   Proposed Method

There are twenty – seven digital fundus images collected from online Digital Retinal Images for Vessel Extraction (DRIVE) database meant for a diabetic retinopathy screening program in The Netherlands [25].

The proposed method includes preprocessing, selfish gene segmentation algorithm and post processing stage. The empirical test is based on a case study of cup and disc segmentation from eye fundus image. The preprocessing stage includes of cropping and removing the vernacular. The processing stage is a selfish gene segmentation algorithm. Finally the post processing stage consists of color segmentation analysis using Receiver Operating Characteristic (ROC) method and subsequently the cup-to-disc ratio (CDR) measurement calculation and evaluation. More explanation will be discussed in subsequent sections. The proposed method is shown in Fig. 1.

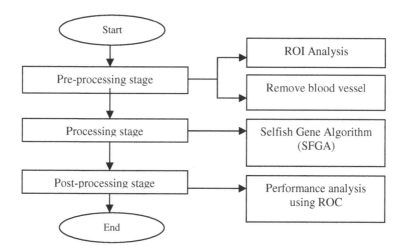

Fig. 1. Flow diagram of the Selfish Gene Algorithm segmentation

2.1   Pre-processing Stage

Pre-processing stage includes two tasks; the region of interest (ROI) and the color channel analysis.

1)   Region of Interest (ROI)

The ROI is centered in the optic cup and disc area as shown in Fig. 2. Due to the high resolution and large size, the image is cropped before it is further processed.s

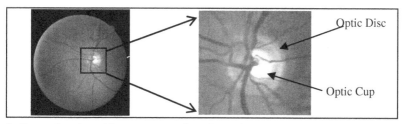

Fig. 2. Cropped region of interest centered on the optic cup and disc from the original fundus image

2) Removing the vernacular

The vernacular is eliminated through the dilation and erosion within the optic disc which also smoothen the intensity profiles around the center of optic disc.

2.2 Processing Stage

Selfish gene algorithm (SFGA) is a new member of Evolutionary Algorithms, search stochastically through a virtual population of the genes, unlike Genetic Algorithm [1,16]. The individual is represented here by its chromosomes. SFGA can be described in 7 steps as below and Fig.7 shows the flowchart of the SFGA.

Encoding representation

In this study, we just focus on the color component as it features in the color fundus image. Color component is divided into 3 channels which are Red (R), Green (G) and Blue (B) as can be seen in Fig. 3. Therefore, these 3 channels are becoming alleles in the SFGA and they have their own frequencies. Each gene is explicitly distinguished between its location in the chromosome (the locus) and the value appearing at that locus (the allele).

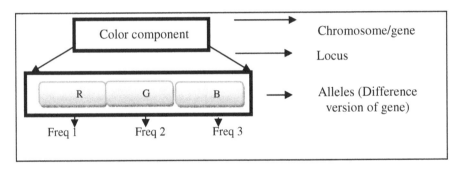

Fig. 3. The illustration of the gene in the Selfish Gene Algorithm (SFGA)

Step 1: Initialize population

A thousand chromosome population is randomly generated from the image. Two chromosomes are evaluated based on predetermined fitness in a tournament. Alleles are determine (group by elements in chromosome) and initialize frequency according to how many times the same alleles occur in population.

Step 2: Evaluate Chromosomes

The winner chromosomes of the tournament will see its genes rewarded. The alleles in the fittest chromosomes are rewarded by increasing its frequencies while the alleles in the non fittest chromosomes penalized by decreasing its frequencies. It can be seen in the Fig. 4 as below.

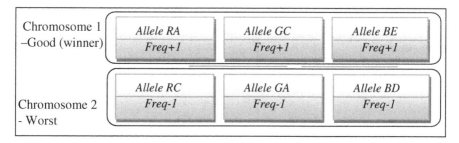

| Chromosome 1 –Good (winner) | Allele RA Freq+1 | Allele GC Freq+1 | Allele BE Freq+1 |
| Chromosome 2 - Worst | Allele RC Freq-1 | Allele GA Freq-1 | Allele BD Freq-1 |

Fig. 4. Example of the chromosome and its fitness for reward and penalized step

Step 3:Update Virtual Population (VP)

The alleles in the chromosomes are separated into alleles population groups based on their locus forming the virtual population (VP) gene pool as shown in Fig.5. Each allele is stored in the form of vector containing the allele value and accumulated frequencies. The accumulated frequencies are obtained following a distribution of probabilities for each gene. The success of an allele is measured by high frequency.

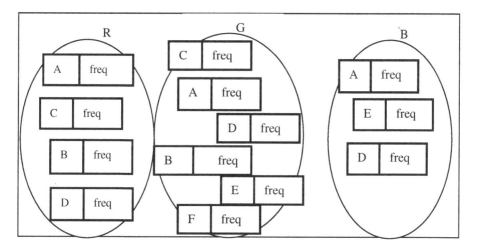

Fig. 5. An example of 4 types alleles in the VP of the SFGA

Step 4: Create Parent

Two parent chromosomes are created based on the combination of highest allele frequency from each VP gene pool according to its locus. Fig.6 shows the example of parent selection from virtual population.

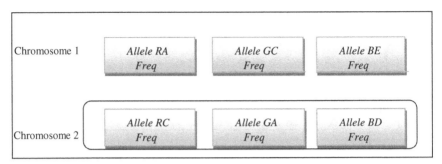

Fig. 6. Example of parent selection from the Virtual Population

Step 5*:* Recombination.

The recombination is done similar to the genetic algorithm method. In this study, one point crossover is used.

Step 6*:* Mutation.

In each competition, genes are randomly chosen and compared. The winner has the opportunity to reproduce.

Step 7*:* Stopping Criteria

Repeat step 2 until a limited number of generations is reached until the required solution found or other needed criteria is met.

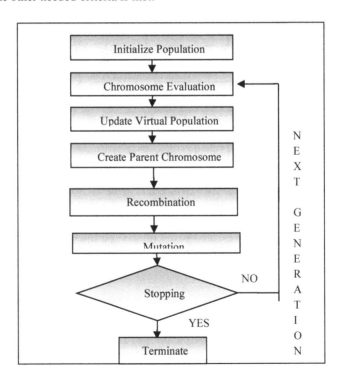

Fig. 7. Flowchart of Selfish Gene Algorithm (SFGA)

2.3   Post-processing Stage

Receiver operating characteristic (ROC) is used to measure the value of false positive (FP), false negative (FN), true positive (TP) and true negative (TN) of the segmented image as shown in Fig. 8. ROC analysis is based between the TP fraction and TN fraction, produced by classifying each pixel as positive and negative from the segmented fundus image.

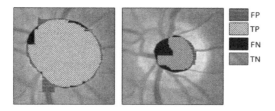

Fig. 8. ROC labeled segmented disc and cup from fundus image

The four primary conditions are used to identify the SFGA segmentation quality and the level of accuracies for optic cup and optic disc segmentation respectively. TN is the normal areas that are correctly undetected while TP is the abnormalities areas that are correctly detected. Meanwhile, FN is the abnormalities areas that are not detected at all and FP is the normal areas that are incorrectly detected as abnormalities area. The description and states for each condition are explained in Table 1 where SA is segmented area, OA is an objective area (optic cup and optic disc) and BS is the background size.

Table 1. Primary condition of ROC analysis

ROC		Description
True Positive	TP	Region segmented as Disc/Cup that proved to be Disc/Cup
True Negative	TN	Region segmented as not Disc/Cup that proved to be not Disc/Cup
False Positive	FP	Region segmented as Disc/Cup that proved to be not Disc/Cup.
False Negative	FN	Region segmented as not Disc/Cup that proved to be Disc/Cup.

Once ROC conditions are determined, the performance of SFGA is measured with sensitivity, specificity and accuracy.

$$\text{Sensitivity} \qquad \frac{TP}{TP+TN} \times 100 \qquad\qquad (1)$$

$$\text{Specificity} \qquad \frac{TN}{TP+TN} \times 100 \qquad\qquad (2)$$

$$\text{Accuracy} \qquad \frac{TP+TN}{TP+FP+FN+TN} \times 100 \qquad\qquad (3)$$

Cup-to-disc ratio(CDR) calculation is derived using Equation 4. The evaluation is based on the value of CDR is greater than 0.3 is considered as glaucomatous eyes while otherwise indicates normal eyes [27]. The CDR ground truth value is determined by the ophthalmologists.

$$CDR = \frac{Height\, of\, Optic\, Cup}{Height\, of\, Optic\, Disc} \tag{4}$$

The segmented image is evaluated based on the ground truth conducted by three ophthalmologists as depicted in Table 2.

Table 2. Segmentation results for SFGA

Feature	Optic Cup	Optic Disc
Ground Truth		
SFGA with Morphological		

3   Results and Discussion

At this stage, the ROC and performance analysis and CDR are evaluated. The proposed method for segmenting the optic cup and disc was evaluated on the basis of the manual outlines drawn by the ophthalmologists. Table 2 shows that the performance of optic disc is better than the optic cup indicated with the higher TP and TN values as depicted in Table 3 The optic cup contour is difficult to measured due to low visibility of the boundary between optic cup and optic rim [26]

Table 3. Summary of ROC for SFGA

Feature	Mean TP	Mean TN	Mean FP	Mean FN
Optic Cup	0.801	0.995	0.005	0.199
Optic Disc	0.857	0.997	0.003	0.143

Table 4. Summary of performance analysis for SFGA

Feature	Sensitivity %	Specificity %	Accuracy %
Optic Cup	80.10	100	90.05
Optic Disc	85.70	100	92.85

4   Conclusion

The new bio-inspired algorithm, the Selfish Gene Algorithm (SFGA) has been presented. Analysis of the literature shows that this algorithm has sparing been applied in other field besides image processing field. Given the characteristics of low image and the preliminary design of the SFGA algorithm, the accuracy of greater than 90% especially with the optic cup shows high potential for the algorithms to be further modified to produce higher accuracy. The selfish gene algorithm has the unique ability of replacing real population with a virtual population (VP). VP stored information about the alleles's frequency in the population for all alleles on all loci. In Virtual Population, the number of individuals, and their specific identity, are not of interest, and therefore are not specified or stored. The first population is created by generating individuals randomly. Individuals are created only for the fitness evaluation and then will be destroyed. When it is successfully fitted, its genes are rewarded. It can be seen when the their frequency in the VP increase. Vice versa, alleles in unfit individuals will be penalized through the reduction in its frequency. Recombination operation is not required since individuals have no role other than for fitness evaluation and the information needed temporarily. The Selfish gene algorithm also is able to achieve higher fitness faster through the independent evolution of the alleles. Thus, giving better solution that other methods. This method has the potential to be more robust and could easily be modified to segment more complexed images such as fine textured images.

References

1. António, C.C.: A memetic algorithm based on multiple learning procedures for global optimal design of composite structures. Memetic Computing 6(2), 113–131 (2014)
2. António, C.C.: Selfish Gene theory and Memetic Algorithms: A fusion of concepts for robust design of hybrid composites (2012)
3. Corno, F., Sonza Reorda, M., Squillero, G.: The Selfish Gene Algorithm: a New Evolutionary Optimization Strategy. In: SAC 1998: 13th Annual ACM Symposium on Applied Computing, Atlanta, Georgia (USA), pp. 349–355 (February 1998)
4. Corno, F., Sonza Reorda, M., Squillero, G.: A New Evolutionary Algorithm Inspired by the Selfish Gene Theory. In: IEEE International Conference on Evolutionary Computation, pp. 575–580 (1998)
5. Corno, F., Sonza Reorda, M., Squillero, G.: Optimizing Deceptive Functions with the SG-Clans Algorithm. IEEE (1999)
6. Corno, F., Sonza Reorda, M., Squillero, G.: Exploiting the Selfish Gene Algorithm for Evolving Hardware Cellular Automata. In: IJCNN2000: IEEE-INNS-ENNS International Joint Conference Neural Networks, Como (Italy), pp. 577–581 (July 2000)
7. Corno, F., Sonza Reorda, M., Squillero, G.: Exploiting the Selfish Gene Algorithm for Evolving Cellular Automata. In: IJCNN2000: IEEE-INNS-ENNS International Joint Conference Neural Networks, Como (I), pp. 577–581 (July 2000)
8. Corno, F., Sonza Reorda, M., Squillero, G.: Exploiting the Selfish Gene Algorithm for Evolving Hardware Cellular Automata. In: CEC2000: Congress on Evolutionary Computation, San Diego, USA, pp. 1401–1406 (July 2000)

9. Corno, F., Sonza Reorda, M., Squillero, G.: Evolving Effective CA/CSTP BIST Architectures for Sequential Circuits. In: SAC2001: ACM Symposium on Applied Computing, Las Vegas, USA, pp. 345–350 (March 2001)

10. Pignalberi, G., Cucchiara, R., Cinque, L., Levialdi, S.: Tuning range segmentation by genetic algorithm. EURASIP Journal Appl. Sig. Proc. 8, 780–790 (2003)

11. Huang, C.F., Rocha, L.M.: A systematic study of genetic algorithms with genotype editing. In: Proc. of 2004 Genetic and Evolutionary Computation Conference, vol. 1, pp. 1233–1245 (2004)

12. Sharma, M.: Memetic Algorithm with Hybrid Mutation Operator. International Journal of Computer Science and Mobile Computing (2014)

13. Lai, C.C., Tseng, D.C.: A hybrid approach using Gaussian smoothing and Genetic algorithm for multilevel thresholding. International Journal of Hybrid Intelligent Systems 1(3), 143–152 (2004)

14. Cao, L., Bao, P., Shi, Z.: The strongest schema learning GA and its application to multilevel thresholding. Image and Vision Computing 26(5), 716–724 (2008)

15. Dawkins, R.: The selfish gene. Oxford University Press, Oxford (1989)

16. Staddon, J.E.R.: Adaptive Behaviour and Learning By J. E. R. Staddon, books (1983)

17. El-Mihoub, T.A., Hopgood, A.A., Nolle, L., Battersby, A.: Hybrid Genetic Algorithms: A Review. Engineering Letters 13(2), EL_13_2_11 (2006)

18. Clow, B., White, T.: An evolutionary race: A comparison of genetic algorithms and particle swarm optimization for training neural networks. In: Proceedings of the International Conference on Artificial Intelligence, IC-AI 2004, vol. 2, pp. 582–588. CSREA Press (2004)

19. Wang, F., Lin, Z., Yang, C., Li, Y.: Using selfish gene theory to construct mutual information and entropy based clusters for bivariate optimizations. Soft Computing 15(5), 907–915 (2011)

20. Ohnishi, K., Koppen, M., Chang, W.A., Yoshida, K.: Genetic clustering based on segregation distortion caused by selfish genes. In: 2012 IEEE International Conference on Systems, Man, and Cybernetics (SMC). IEEE (2012)

21. Ghanea-Hercock, R.: Applied Evolutionary Algorithms in Java. Springer-Verlag New York, Inc. (2003) ISBN 0-387-95568-2

22. Blickle, T.: Theory of evolutionary algorithms and application to system-synthesis. Ph.D. dissertation, Swiss Federal Inst. Technol (ETH), Zurich, Switzerland, 1996, ETH diss no. 11894 (1996)

23. Hu, J.: Sustainable Evolutionary Algorithms And Scalable Evolutionary Synthesis Of Dynamic Systems, PhD thesis, Department of Computer Science and Engineering, Michigan State University, East Lansing, Michigan, 48823, USA, 2004. Erik Goodman, Advisor (2004)

24. Espejo, P.G., Ventura, S., Herrera, F.: A Survey on the Application of Genetic Programming to Classification. IEEE Transactions on Systems, Man, and Cybernetics, Part C, 121–144 (2010)

25. ISI.: Image Database at Image Sciences Institute. Retrieve from I age Science Institutes (2001)

26. Liu, J., Yin, F.S., Wong, D.W.K., Zha, Z., Tan, N.M., Cheung, C.Y., Baskara, M., Aung, T., Wong, T.Y.: Automatic Glaucoma Diagnosis from Fundus Image. In: Engineering in Medicine and Biology Society, EMBC, 2011 Annual International Conference of the IEEE Page (s): 3383 – 3386 (2011)

27. Jagadish, N., Rajendra, A.U., Subbanna, P.B., Nakul, S., Teik, C.L.: Automated Diagnosis of Glaucoma Using Digital Fundus Images. J. Med. Syst. 33, 337–346 (2009)

Detecting IMSI-Catcher Using Soft Computing

Thanh van Do[1,2], Hai Thanh Nguyen[1], Nikolov Momchil[1], and Van Thuan Do[3]

[1] Telenor ASA. Snarøyveien 30 1331 Fornebu, Norway
[2] Norwegian University of Science and Technology,
O.S. Bragstadsplass 2B 7031 Trondheim, Norway
[3] Linus AS, Martin Linges vei 15, 1364 Fornebu, Norway
{thanh-van.do,haithanh.nguyen}@telenor.com,
Mnikolov@telenor.bg, t.do@linus.no

Abstract. Lately, from a secure system providing adequate user's protection of confidentiality and privacy, the mobile communication has been degraded to be a less trustful one due to the revelation of IMSI catchers that enable mobile phone tapping. To fight against these illegal infringements there are a lot of activities aiming at detecting these IMSI catchers. However, so far the existing solutions are only device-based and intended for the users in their self-protection. This paper presents an innovative network-based IMSI catcher solution that makes use of machine learning techniques. After giving a brief description of the IMSI catcher the paper identifies the attributes of the IMSI catcher anomaly. The challenges that the proposed system has to surmount are also explained. Last but least, the overall architecture of the proposed Machine Learning based IMSI catcher Detection system is described thoroughly.

Keywords: IMSI catcher detection, mobile phone tapping, phone eavesdropping, machine learning, anomaly detection.

1   Introduction

Until recently, mobile communication has been perceived by the majority of users as quite secure regarding both confidentiality and privacy thanks to the strong encryption combined with use of temporary identities. In fact, users quite often consider mobile telephony as more secure than fixed telephony. Recently, a series of scandalous phone tapping incidents in the United States, United Kingdom, Germany, China, etc. revealed by Snowden, a former American National Security Agency (NSA) agent had eroded this conviction. It is really shocking that not only very important people at high position like the German chancellor, prime ministers, members of parliament, etc. but also regular people may be victims of phone eavesdropping. But most frightening lies perhaps in the fact that the monitoring may be done by anybody from the police, governmental intelligence agencies, security institutions, etc. to private companies or organisations. With advances in microelectronics and the availability of mobile open source software, equipment used in phone tapping are getting both smaller, easier to handle, more available and also quite affordable in the range of US $1500-2000.

© Springer Science+Business Media Singapore 2015
M.W. Berry et al. (Eds.): SCDS 2015, CCIS 545, pp. 129–140, 2015.
DOI: 10.1007/978-981-287-936-3_13

Since the last couple of months, Aftenposten [1], one of the biggest newspapers in Norway has published several articles telling that they have detected the presence of many IMSI catchers aka fake base stations in the region of Oslo that could be used in the surveillance of mobile users.

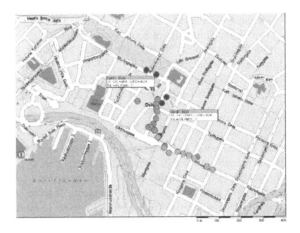

Fig. 1. IMSI catchers detected around the Parliament in Oslo (courtesy: Aftenposten)

The detection has been done using mobile devices such as GSMK Cryptophone [2] that the reporters carried with them when moving around in Oslo. In fact, the existing IMSI detection solutions are based on portable devices that monitor the radio access network to detect possible presence of IMSI catchers. There is today no network based solution to detect IMSI catchers and the reasons are twofold. First, the need for IMSI detection is so far non-existent because there are only a few IMSI catchers used by governmental agencies in the fight against crimes and terrorism. Secondly, mobile operators do not consider IMSI catchers as threats because they just monitor the users' conversation and do not do any harm to their mobile networks. However, with the increasing number of mobile phone tapping incidents, the users start to lose confident in mobile communication and mobile operators begin to realise that something must be done. This paper introduces a Machine Learning based IMSI catcher Detection system that was initiated by the Telenor ImobSec project in collaboration with Norwegian universities and security experts. The paper begins with a review of related works. Next, a comprehensive explanation of the IMSI catcher is given. For the detection of IMSI catcher the attributes of the anomaly input data set are then identified and clarified. The challenges to the proposed system are also analysed. The central part of the paper is the proposed Machine Learning based IMSI catcher Detection system which is described thoroughly. Further works are proposed in the conclusion.

2   Related Works

There are currently several initiatives aiming at developing IMSI detection solutions and as the review in this section will show they are all mobile device based solutions.

2.1   Security Research Labs (SRLabs)

The SRLabs [3] in Berlin directed by the famous German Cryptographer and security scientist Karsten Nohl has conducted a few activities focused on the detection of mobile phone tapping. For the assessment of mobile network security SRLabs offers a set of tools.

CatcherCatcher
One of these tools is the CatcherCatcher tool which has the ability to detect mobile network irregularities suggestin a fake base station activity. The CatcherCatcher consists of:

* Osmocom₁ phone [4]
* Osmocom cable
* Linux computer

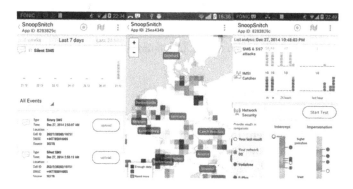

Fig. 2. SnoopSnitch main Views

SnoopSnitch
SnoopSnitch is an Android application which by collecting and analysing wireless radio data, makes users aware of their mobile network security and warns them about possible threats such as fake base stations (IMSI catchers), user tracking and Over The Air updates as shown in Figure 2. With SnoopSnitch users can both make use and contribute data to the GSM Security Map at gsmmap.org.

---

¹ The Osmocom project is a family of projects that are related Open source mobile communications. Its provides software and tools for a variety of mobile communication standards, including GSM, DECT, TETRA and others.

This application is currently working only on Android phones with a Qualcomm chipset and a stock Android ROM (or a suitable custom ROM with Qualcomm DIAG driver). Root privilege is required to capture mobile network data.

2.2   Android IMSI-Catcher Detector (#AIMSICD)

#AIMSICD [5] is an Android-based project originated from XDA forum[2] with the goal of detecting and avoiding fake base stations (IMSI-Catchers) in GSM/UMTS networks which receives contributions from a large amount of anonymous and public Android developers, baseband hackers. All the developments are fully open source under GPL v3+ and located in an official GitHub repository.

Fig. 3. #AIMSICD main views: Main Screen, Cell Information, Database Viewer, Map Viewer (Courtesy: #AIMSICD)

The project has achieved the following:

- Main Views as shown in Figure 3:
  - o **Main Screen:** Information about Device, Network and SIM-Card
  - o **Cell Information:** Relevant variables using public AOS API calls (LAC, CID, Signal Strength)
  - o **Database Viewer:** Data collected by the phone and from public DB of Cell Towers
  - o **Map Viewer:** Shows Cell-Towers from the public database that are in your area
- Functions:
  - o **Cell-Monitoring:** Collects information about the cell towers you are/were connected to and saves it in the local database
  - o **Cell Tracking:** Tracks your position while you are connected to the cell-tower and saves it with cell data in the local database
  - o **Download data** from public cell-tower database (right now only from OpenCell_ID)

---

2 XDA Developer (also known simply as XDA; often denoted as xda-developers) is a mobile software development community of over 5 million members worldwide, started in January 2003.

- o   **Position Tracking:** Using the GPS-Sensor and Google Location Service
- • Detection:
  - o   **Check Cell_ID's** collected by the phone against public Cell-Tower Database.
  - o   **Check for "Changing LAC"** of each Cell-ID that is collected by the phone.

2.3   SBA Research

SBA Research is an Austrian research center for Information Security funded by the national initiative for COMET Competence Centers for Excellent Technologies. It enables the collaboration of 25 companies, 4 Austrian universities, one university of applied sciences, a non-university research institute, and many international research partners on challenges categorizing from organizational to technical security. SBA Research has two independent implementations of an IMSI Catcher Catcher (ICC). The first one employs a network of stationary (sICC) measurement units installed in a geographical area and constantly scanning all frequency bands for cell announcements and fingerprinting the cell network parameters. These rooftop-mounted devices can cover large areas. The second implementation is an app for standard consumer grade mobile phones (mICC), without the need to root or jailbreak them [6].

3   Brief Description of IMSI Catcher

An IMSI catcher is a device for intercepting GSM mobile phones. It subjects the phones in its vicinity to a Man-In-The-Middle (MITM) attack by pretending to be the preferred base station in terms of signal strength.

As its name tells, the IMSI catcher logs the IMSI numbers of all the mobile phones in the area, as they attempt to attach to the base station, and can determine the phone number of each individual phone. It also allows forcing the mobile phone connected to it to revert to A5/0 for call encryption (in other words, no encryption at all), making the call data easy to intercept and convert to audio. The phone calls can hence be tapped and recorded by the IMSI Catcher.

This specific MITM attack was patented by Rohde & Schwarz, which presented the first IMSI Catcher GA 090 in Munich in 1996. On 24 January 2012, the Court of Appeal of England and Wales held that the patent is invalid for obviousness, since in reality it is just a modified cell tower with a malicious operator.

The GSM specification requires the handset to authenticate to the network, but does not require the network to authenticate to the handset. This is obviously a weakness but GSM is a 26 years old technology and at its specification time it was almost impossible to have access to a false base station and a mutual authentication would be too heavy for the SIM card. IMSI catchers are employed by law enforcement and intelligence agencies.

To remedy the weakness of GSM, UMTS (3G) and LTE (4G) introduce mutual authentication, which requires also the authentication of base stations towards the mobile handset. Unfortunately, due to backward compatibility and the use of GSM as a fallback network where UMTS is not available, mobile phones can be forced to downgrade to a 2G connection and fully exposed to tapping [6].

Fig. 4. Various current IMSI catchers (courtesy: #AIMSICD)

As shown in Figure 4, not like the first generation IMSI catchers which were big, heavy and expensive, the current ones come in uncountable shapes, sizes and prices, and can be as tiny as the portable Septier IMSI-Catcher Mini.

4   Detection of IMSI Catcher

To detect the presence of IMSI catcher it is necessary identify the anomalies in the mobile networks and to define the nature of anomaly detection input data set [7]. The outlier (anomaly) detection approach type 3 [8] is chosen since we already have a lot of knowledge about the mobile network. A few cases of IMSI catcher presence are considered to model anomalies and to define the input data set.

4.1   Camping in 2G Instead of 3G

In many areas, especially the urban ones, 2G, 3G and 4G networks will quite often coexist to accommodate all kinds of handsets and subscriptions. A 3G enabled handset will normally connect itself to 3G networks since services with higher QoS could be provided. An IMSI catcher would jam the signals of the 3G base stations and force the mobile phone to disconnect from the original network and register to it. The IMSI catcher [9] acts as a base station toward the mobile station and as a mobile station toward the real base station as shown in Figure 5. The IMSI catcher can establish a regular connection with the mobile network using its own SIM or without a SIM. In the latter case, in the authentication process, the IMSI catcher simply forwards the authentication data received from the mobile terminal to the base station and conveys the data from the base station to the mobile phone. The session has to be conducted with disabled encryption in both ways since the IMSI catcher does possess the ciphering key K_c.

Seen from the network anomalies created by the IMSI catcher in an area may have as a contextual attributes the *percentage of mobile phones on 2G* and the *one of mobile phones on 3G*. Indeed, a 2G percentage higher than average or a 3G percentage lower than average will be considered as contextual anomalies. Another contextual attribute is the *percentage of 3G enabled mobile handset* that together with the two previous ones constitute a collective anomaly. The anomaly model can be improved further by the addition of more attributes such as *signal strength, antenna type* (omnidirectional, sectorial), *cell ID, cell size*, etc. such that the detection can be improved and false positive reduced.

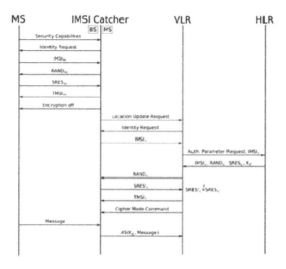

Fig. 5. Man-in-the-middle attack with an IMSI Catcher

4.2   Temporary Disappearance of Mobile Phones

Quite often IMSI catchers have to jam all the signals in its vicinity. This will make the mobile stations around lose their connections to the mobile network. They will not be reached by the paging of the base station controller (BSC) at incoming calls addressed to them. The calls can hence not be delivered to the mobile phones as if they have moved to an area without coverage or if there are switched off or ran out of power. Anyway they seem to disappear from the network.

The *number of mobile phones* that disappear for certain period of time from a network area will be used as a condition attribute for the IMSI catcher anomaly because a high value of disappeared mobile phones can indicate the possible presence of an IMSI catcher.

4.3   Disabling of Encryption

Although not all the IMSI catchers the ones that do not use their own SIM to establish a connection to the mobile network, must disable encryption to tap the call since they do not have access to the ciphering key K_c. Although they have lower probability to succeed due the security requirement at many mobile operators these IMSI catchers are still in operation in many countries.

The number of call established with disable encryption can be used a contextual attribute for the IMSI catcher anomaly because a high value of calls with disable encryption can be an indication of the presence of an IMSI catcher.

4.4   Challenges to the IMSI Catcher Anomaly

Although using anomaly detection in the detection of IMSI catcher there are, however, a few challenges that need to be considered carefully as follows:

Determination of the Area for the IMSI Catcher Anomaly
To determine the area for the collection of the input data set could be a challenging task because 2G (GSM), 3G (UMTS) and 4G (LTE) have different partitioning for location updating of the mobile phone.

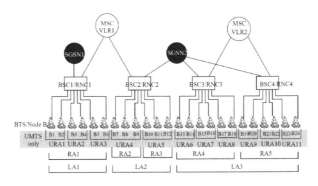

Fig. 6. Geographical partitioning in UMTS

A GSM network (2G) is divided into cells, which are grouped into a *location Area (LA)*. A mobile phone in motion keeps the network informed about changes of the location area [10]. When moving from a cell belonging to a location area to another cell belonging to another location area, the mobile terminal has to perform a location area update to inform the network about the new location area in which it is located. A location area is usually managed by a base station controller (BSC) and could be a candidate for the data collection area.

In a UMTS network (3G) a serving GPRS support node (SGSN) manages one or more Radio Network Controllers (RNC) and an RNC manages several Node B (equivalent to base station in GSM) [11].

To track the location of a UE (mobile phone) some geographical groups are defined within the UMTS Radio Access Network (UTRAN) as follows [12]:

- *Location Area (LA):* covers the area of one or more Radio Network Controllers (RNC) managed by the same SGSN.
- *Routing Area (RA)*: is a subset of a LA. It only covers one RNC or even only a subset of an RNC.

- *UTRAN RA (URA):* is a subset of an RA. It only covers multiple NodeBs of one RNC.

LAs are used in the Circuit-Switched-domain and RAs in the Packet-Switched domain. As shown in Figure 6 Geographical partitioning in UMTS, a LA can have one or more RNC and one or more RA. An RA can have one or more URA. All the geographical partition may be considered as data collection area since the mobile station could be tracked in all of them.

In an LTE network (4G) the cells (eNodeB) are grouped in *Tracking Area (TA)* used for paging of the UE (Mobile phone) [13]. Tracking areas can be grouped into lists of Tracking Areas (TA lists), which are administered by the User Equipment (UE). Tracking area updates are performed periodically or when the UE moves to a tracking area that is not included in its TA list. Mobile operators can allocate different TA lists to different UEs. By this way signaling peaks can be avoided in some conditions. For example, User equipment of train passengers may not perform tracking area updates simultaneously. The dimension of the TA compared to the LA depends on many factors like LTE paging capacity, TA update overhead, LA update overhead, etc. and could vary depending on the network. The Tracking Area (TA) may be considered as a data collection area because the mobile phone could be tracked in it.

At the first glance, the Location Area of 2G is most suitable to be the data collection area but considerations have to be done when performing the mapping of Routing Area (RA), UTRAN (URA) and Tracking Area (TA) to LA because full match may not be achieved.

Collection of Data

The attributes of the input data set for the IMSI catcher are operational data of the mobile networks. They are collected, used and disposed within a period of time because they are huge in amount. For a more permanent storage of these attributes upgrades on the mobile networks have to done and this could pose financial issues.

The ultimate goal of the IMSI catcher detection system is to identify the fake base stations as quickly as possible. Since those base stations quite often are on the move the collection of data has to be done mostly in real time. This becomes a bigger challenge when the input data are collected from distributed network components. A solution could be a distributed system having a collection and detection function for each data collection area e.g. Location Area.

5 The Proposed Machine Learning Based IMSI Catcher Detection System

To utilize and combine the previously mentioned contextual attributes for the detection of IMSI catcher we propose an innovative detection system based on machine-learning techniques which consists of two main parts as shown in Figure 7 namely the online-detection part and the off-line learning part.

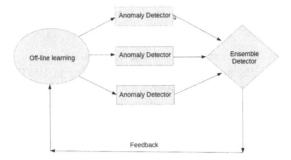

Fig. 7. The proposed Machine Learning based IMSI catcher Detection system

As shown in Figure 7, the online-detection part contains different anomaly detectors, each of which uses a contextual attribute to define normal or abnormal behavior. For example, in our case we can have three different anomaly detectors. The first one is based on the change between 2G and 3G modes. The second one takes into account the temporary disappearance of mobile phones. The third one uses the disabling of the encryption. To combine these detectors, an ensemble detector is needed. A simplest form of the ensemble model is the majority voting between the different detectors but a weighted voting may also be considered in later phases. Several machine-learning algorithms, such as one-class Support Vector Machines [14] and Neural Networks, can be used as anomaly detectors.

Following the suggestions from the ensemble detector, security experts would then look at suspicious places to verify if there any true IMSI catchers at a point in time. The feedback from the security experts is then given back to off-line learning part to update the models where the normal behavior was defined.

At the present stage of our project no real data set has been yet collected from the mobile network and for illustration sake we did some experiments on the public available data set related to IMSI catcher detection from Aftenposten [15]. The data came from a handset while interacting with the mobile network and possibly with the IMSI catchers as well, and we cannot show the advantages of machine-learning in correlating different events from many different devices. However, the main objective of this experiment is to show that there is a potential of applying machine learning techniques to facilitate the detection process.

For our simple experiment, from the data set we were interested in the frequency of the mode change between 2G and 3G. Our hypothesis is that the high value of the frequency would indicate abnormality in an area.

We split the data by equal time slots and calculate the ratio between the number of 2G and 3G in each time slot. We applied the anomaly detection algorithm named S-H-ESD from Twitter [16] to detect abnormalities for those obtained ratio values. The result is shown in Figure 8. The three high spikes, which denote the abnormalities, indicate the possible presence of IMSI catcher.

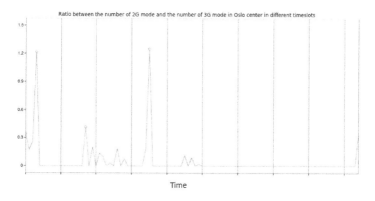

Fig. 8. Experiments on 2G/3G modes with Aftenposten data set

6   Conclusion

In this paper an innovative system for the detection of IMSI catcher is presented and described. It has a few advantages compared to the current existing solutions. First, according to our knowledge, most of the current IMSI catcher detectors, are located on the handset side, either as a dedicated device or an application that can be downloaded and installed in a regular mobile phone. They are intended for the users and put the responsibility of protecting confidentiality and privacy on the users' shoulders. This may be both unfair and unmanageable for non-technical users. The proposed Machine Learning based IMSI catcher Detection system is network based and intended for mobile operators in the protection of their users. It is offering a more balanced and fair solution. Second, by being network based the proposed system will be able to carry out the data collection and detection at several areas simultaneously and hence improving the ability to detect an IMSI catcher. Third, the proposed detection system is using machine learning techniques and can hence learn and enhances itself more rapidly than the current IMSI catcher. However, as stated in the paper, there are also a few challenges such as determination of the area for the IMSI catcher anomaly, and collection of data, that have to be surmounted. The ImobSec project is also in very earlier phase and only data set from Aftenposten has been experimented. As future works, the proposed Machine Learning based IMSI catcher Detection system will be installed and deployed in the Telenor Norway mobile test network. Experiments will then be carried out with the introduction of the project's IMSI catcher in the network. The lessons learned will then be applied to optimize and improve the system further.

References

1. Foss, A.B., Johansen, P.A., Hager-Thoresen, F.: Secret surveillance of Norway's leaders detected; Aftenposten (December 16, 2104).
   http://www.aftenposten.no/nyheter/iriks/Secret-surveillance-of-Norways-leaders-detected-7825278.html

2. GSMK CRYPTOPHONE: `http://www.cryptophone.de/en/`
3. Security Research Labs: `https://opensource.srlabs.de/`
4. The Osmocom (Open Source Mobile Communication) project: `http://openbsc.osmocom.org/trac/wiki/OsmocomOverview`
5. Android IMSI-Catcher Detector (#AIMSICD); `https://secupwn.github.io/Android-IMSI-Catcher-Detector/`
6. Dabrowski, A., Pianta, N., Klepp, T., Mulazzani, M., Weippl, E.R.: IMSI-Catch Me If You Can: IMSI-Catcher-Catchers. In: Annual Computer Security Applications Conference (ACSAC). ACM (2014); 978-1-4503-3005-3/14/12
7. Chandola, V., Banerjee, A., Kumar, V.: Anomaly detection: A survey. ACM Computing Surveys 41(3), 1 (2009), doi:10.1145/1541880.1541882
8. Hodge, V.J., Austin, J.: A Survey of Outlier Detection Methodologies. Artificial Intelligence Review 22(2), 85 (2004), doi:10.1007/s10462-004-4304-y
9. Strobel, D.: IMSI Catcher, Chair for Communication Security, Ruhr-Universität Bochum (July 13, 2007)
10. 3GPP: Technical Specification TS 23.012 Location management procedures, V12.0.0 (September 2014)
11. 3GPP: Technical Specification TS 23.060 GPRS Service Description describes, V13.2.0 (March 2015)
12. Chen, Y.-S.: Chapter 2 Mobility Management for GPRS and UMTS; Department of Computer Science and Information Engineering National Taipei University
13. ElNashar, A., El-saidny, M., Sherif, M.: Design, Deployment and Performance of 4G-LTE Networks: A Practical Approach. John Wiley & Sons (2014); ISBN 1118703448, 9781118703441
14. Vert, R., Vert, J.-P.: Consistency and Convergence Rates of One-Class SVMs and Related Algorithms. JMLR 7, 817–854 (2006)
15. Aftenposten data set. `http://www.aftenposten.no/meninger/kommentarer/Derfor-publiserer-Aftenposten-hele-datagrunnlaget-for-mobilspionasje-sakene-7849555.html`
16. Anomaly Detection algorithm from Twitter. `https://github.com/twitter/AnomalyDetection`

Solving Curriculum Based Course Timetabling by Hybridizing Local Search Based Method within Harmony Search Algorithm

Juliana Wahid[1] and Naimah Mohd Hussin[2]

[1] School of Computing, College of Arts and Sciences, Universiti Utara Malaysia,
06010 UUM Sintok Kedah, Malaysia
w.juliana@uum.edu.my
[2] Faculty of Computer Science and Mathematical,
Universiti Teknologi MARA (Perlis), 02600 Arau Perlis, Malaysia
naimahmh@perlis.uitm.edu.my

Abstract. The curriculum-based university course timetabling which has been established as non-deterministic polynomial problem involves the allocation of timeslots and rooms for a set of courses depend on the hard or soft constraints that are listed by the university. To solve the problem, firstly a set of hard constraints were fulfilled in order to obtain a feasible solution. Secondly, the soft constraints were fulfilled as much as possible. In this paper we focused to satisfy the soft constraints using a hybridization of harmony search with a great deluge. Harmony search comprised of two main operators such as memory consideration and random consideration operator. The hybridization consisted three setups based on the application of great deluge on the operators of the harmony search. The great deluge was applied either on the memory consideration operator, or random consideration operator or both operators together. In addition, several harmony memory consideration rates were applied on those setups. The algorithms of all setups were tested on curriculum-based datasets taken from the International Timetabling Competition, ITC2007. The results demonstrated that our approach was able to produce comparable solutions (with lower penalties on several data instances) when compared to other techniques from the literature.

Keywords: Harmony Search, Great Deluge, Curriculum Based Course Timetabling

1 Introduction

Curriculum-based course timetabling (CBCTT) is considered as non-deterministic polynomial (NP) problem that is intractable, i.e. there is no efficient algorithm that is guaranteed to find an optimal solution for such problems [1]. This is parallel to the theory of no-free-lunch theorem [2] which states that if any prior assumptions cannot be made about the optimization problem we are trying to solve, no algorithm can be expected to perform better than any other algorithm on that problem. The design of new methods and techniques to solve CBCTT problem is a very active area of

© Springer Science+Business Media Singapore 2015
M.W. Berry et al. (Eds.): SCDS 2015, CCIS 545, pp. 141–153, 2015.
DOI: 10.1007/978-981-287-936-3_14

research. A very promising research area is the hybridization of metaheuristics techniques [3]. This paper focuses on the hybridization of metaheuristics between the population-based method (harmony search) and local search based method (great deluge). The aim of hybridizing a local search based with a population-based method is to attain a balance between exploration and exploitation of the search space utilizing the advantage of population-based and local search based methods [4], [5].

2   Curriculum Based Course Timetabling Problem

The CBCTT problem is to create a weekly timetable of courses with meeting time by allocating the lectures to a certain number of rooms and timeslots based on the curricula [6]. The CBCTT problem definition consists of the basic entities shown in Table 1. The entities are described in the second column.

Table 1. Basic entities in curriculum based course timetabling.

Entity	Definition
Days (d)	Number of teaching days in the week (typically 5 or 6).
Timeslots (ts)	Each day is split into a fixed number of timeslots, which is equal for all days.
Periods (P) = d X ts	A pair composed of a day and a timeslot. The total number of scheduling periods is the product of the days times the day timeslots. A set of P periods, T=$\{T_1,...T_P\}$.
Courses and Teachers	A set of N courses, C = $\{C_1,...,C_N\}$, each course is composed of the number of lectures (L), to be scheduled and each lecture is associated to a teacher.
Rooms	Each room has a capacity, expressed in terms of number of available seats (c), and a location expressed as an integer value representing a separate building (l). Some rooms may not be suitable for some courses (because they miss some equipment). A set of M rooms, R=$\{R_1,...R_M\}$.
Curricula	A curriculum is a group of courses such that any pair of courses in the group have students in common. Based on curricula, we have the conflicts between courses and other soft constraints. Set of Q curricula Cu = $\{Cu_1, Cu_2, ..., Cu_Q\}$

A feasible timetable is when all lectures have been scheduled for a time slot and a room, so that the hard constraints are satisfied. On the other hand, soft constraints may be violated. Then the objective of CBCTT problem is to minimize the number of soft constraint violations in a feasible solution in order to improve the solution quality.

A previous study [7] has divided the constraint into two sets: the hard constraints and the soft constraints. (1) Hard constraints: H1 - Lectures: All lectures of a course must be scheduled, and they must be assigned to distinct periods. H2 - Conflicts: Lectures in the same curriculum or taught by the same teacher must all be scheduled in different periods. H3 - Room occupancy: Two lectures cannot take place in the same room at the same time. H4 - Availability: If the teacher of the course is not available to teach that course at a given period, then no lecture of the course can be scheduled at that time. (2) Soft constraints: S1 - Room Capacity: For each lecture, the number of students that attend the course must be less or equal than the number of seats in all rooms that host the lectures. S2 - Min Working Days: The lectures of each course must spread into the given minimum number of days. S3 – Isolated Lectures: Lectures that belong to a curriculum should be adjacent to each other (i.e., in consecutive periods). S4 - Room Stability: All lectures of a course should be given the same room. The quality of solution is calculated as the total penalties of the soft constraints: S1 + S2 + S3 + S5.

3   The Algorithm

The algorithm consisted of construction and improvement algorithms.

3.1   Construction Algorithm

Starting from an empty timetable, each of the lectures which were sequenced by saturation degree heuristic, followed by largest degree heuristic and went through a lecture assignment process. A conflict matrix was produced in the pre-processing phase to identify the conflict for each lecture. Sometimes lectures did not have a conflict (in certain data instances) because they existed on their own in the curricula. These lectures were assigned later, after all the lectures with conflict had already been assigned.

Each of the lectures which have already sequenced according to the above heuristics setting was randomly assigned to empty slots in an iterative process. During this process, the feasibility of all hard constraints in each iteration step was enforced. In the case where a lecture cannot be assigned to any slots due to non-available slots, it was then included in the unassigned lectures list.

The unassigned lecture list was further assigned to the timetable using nine procedures that were carried out in sequence. If a feasible timetable is found at any level of procedures, the algorithm keeps the solution and start new assignment procedures using a different random seed. Otherwise, all other procedures are repeated with a maximum of 50 iterations. After 50 iterations of unassigned lecture assignment procedures, if a feasible solution is not found, the current timetable will be discarded and the algorithm start a new assignment procedure using a different random seed. For lectures without a conflict, it was scheduled in the timetable using the same approaches. However, only two procedures were used.

The whole process was repeated 50 times to produce a population of initial solutions as the improvement algorithm used is a population-based method. Only feasible timetable (i.e. with no hard constraints) was included in the population of initial solution.

3.2   Improvement Algorithm

Harmony search. The harmony search algorithm (HSA) is a population based on metaheuristic algorithm that impersonates the musical improvisation process in which a group of musicians improvise their instruments' pitch by searching for a perfect state of harmony according to audio-aesthetic standard [8]. The HSA requires six steps as follows:

1. Determine the algorithm parameter setting - the HSA parameters required to solve the CBCTT problem are harmony memory consideration rate (HMCR), harmony memory size (HMS) (that is equivalent to population size), pitch adjustment rate (PAR), and maximum improvisations (MI) (that is the maximum number of generations)
2. Memory initialization – the process of constructing the population of initial solution which is called harmony memory (HM). The number of initial solution is determined by the value of HMS stated in the first step.
3. Harmony improvisation - the solution is optimized (improved) using the following operators:
   (a) Memory consideration (MC) - choosing the lecture (from the HM) to be assigned in the timetable slot.
   (b) Random consideration (RC) - choosing the lecture from all lectures that are available.
   (c) Pitch adjustment (PA) - replaced the lecture assigned by memory consideration (MC) operator.
4. Update memory with the solution found - the new solution that is better than previous solution is included in HM.
5. Determine the termination criteria - the termination criteria used is the number of the iteration process, i.e. maximum improvisations (MI) that has been defined in the first step.
6. Cadenza (musical terminology) - return the best harmony

Great Deluge. The great deluge (GD) algorithm is proposed by [9] and is motivated by the behaviour of the hiker who seeks the peak of the ground when the water level rises up during rainy season. GD is a variance of simulated annealing (SA) technique in which the differences such as GD involves fewer parameters and decrease the objective function in its acceptance rule of solutions [10].

Hybridization of Harmony Search and Great Deluge. The hybridization of GD within HSA consists of three types: (1) hybridization of HSA and GD in RC operator (NGD), (2) HSA with GD in MC operator (GDN) and (3) HSA with GD in MC and RC operator (GDGD). The N in the NGD and GDN consists of an algorithm as in the original HSA described above.

Figure 1 shows the general pseudo code of the hybridization of HSA and GD, while Table 2 shows the respective MC, RC, and UPDATE acceptance formula related to the hybridization. The NN hybridization is the original HSA implementation without local search based method hybridization.

Step 1: HSA *parameters settings (HMS,HMCR, PAR, MI)*
Step 2: *Initialize HM{x_1,..., x_{HMS}}*
while *not termination criterion specified by MI* **do**
Step 3: *Harmony Improvisation*
    Select the best harmony $x^{BEST} \in (x_1,..., x_{HMS})$.
    Set Current Best harmony, $x^{CURBEST} = x^{BEST}$
    Set water level, B= f(x^{BEST})
    for j =1,..., N **do** *(N is the number of decision variables)*
        *if $U(0,1) \le HMCR$ **(memory consideration)***
            (pitch adjustment)
            Move timeslot: $0 \le U(0,1) \le 0.2xPAR$
            Swap timeslot: $0.2xPAR < U(0,1) \le 0.4xPAR$
            Move room: $0.4xPAR < U(0,1) \le 0.6xPAR$
            Swap room: $0.6xPAR < U(0,1) \le 0.8 xPAR$
            Kempe chain move: $0.8xPAR < U(0,1) \le 1 xPAR$
            MC Acceptance Formula
        end if (end of memory consideration)
        else (random consideration)
            Move or Swap (timeslot and room)
            RC Acceptance Formula
        end if (end of random consideration)
    end for
Step 4: *Update the new harmony in the HM*
    UPDATE Acceptance Formula
end while (Step 5: *Performing termination)*
Step 6: *Cadenza (returns the best harmony ever found*

Fig. 1. General Pseudo Code of HSA and GD Hybridization

In the pseudo code, in step 1, the values of HSA parameters were initialized by the following values: HMS = 50, HMCR = 0.2, 0.5, and 0.8, PAR = 1.0, MI = 1000. In step 2, the initialization of harmony memory was the construction algorithm which already described in section 3.1.

During the improvement stage, i.e. step 3, the best solution, x^{BEST} from the harmony memory (HM) was selected and assigned to current variable, $x^{CURBEST}$. The cost of the best solution $f(x^{BEST})$ was assigned to B which was the initial water level.

For N iterations, the improvisation step in MC operator and RC operator in the hybridization of HSA with GD algorithm constituted several neighborhood structures as follows:

- The move timeslot. With the probability between 0%×PAR and 20%×PAR, the lecture is randomly moved to any feasible timeslot in the same room.
- The swap timeslot. With the probability between 20%×PAR and 40%×PAR, the lecture is swapped with the timeslot of another lecture, while the rooms of both lectures are not changed.

- The move room. With the probability between 40%×PAR and 60%×PAR, the lecture is randomly moved to any feasible timeslot in a different room.
- The swap room. With the probability between 60%×PAR and 80%×PAR, the lecture is swapped with the timeslot of another lecture located in a different room.
- The kempe chain move. With the probability between 80%×PAR and 100%×PAR, the lecture is moved using a Kempe chain. A Kempe chain is defined as a set of lectures that form a connected component because the conflict in the subset of lectures belongs to two distinct periods.

Table 2. Different Acceptance formula in hybridization of HSA and GD.

Acceptance Formula			Hybridization's name
MC	**RC**	**UPDATE**	
if $f(x^{NEW}) <= f(x^{CURBEST})$ $x^{CURBEST} = x^{NEW}$ *end if*	*if* $f(x^{NEW}) <= f(x^{CURBEST})$ $x^{CURBEST} = x^{NEW}$ *end if*	*if* $f(x^{NEW}) <= f(x^{WORST})$ $x^{WORST} = x^{NEW}$ *end if*	*NN*
if $f(x^{NEW}) <= f(x^{CURBEST})$ $x^{CURBEST} = x^{NEW}$ *end if*	*If* $f(x^{NEW}) <= B$ *or* $f(x^{NEW}) <= f(x^{CURBEST})$ $x^{CURBEST} = x^{NEW}$ *end if*	*If* $f(x^{NEW}) <= B$ *or* $f(x^{NEW}) <= f(x^{WORST})$ $x^{WORST} = x^{NEW}$ *end if*	*NGD*
If $f(x^{NEW}) <= B$ *or* $f(x^{NEW}) <= f(x^{CURBEST})$ $x^{CURBEST} = x^{NEW}$ *end if*	*if* $f(x^{NEW}) <= f(x^{CURBEST})$ $x^{CURBEST} = x^{NEW}$ *end if*	*if* $f(x^{NEW}) <= B$ *or* $f(x^{NEW}) <= f(x^{WORST})$ $x^{WORST} = x^{NEW}$ *end if*	*GDN*
If $f(x^{NEW}) <= B$ *or* $f(x^{NEW}) <= f(x^{CURBEST})$ $x^{CURBEST} = x^{NEW}$ *end if*	*If* $f(x^{NEW}) <= B$ *or* $f(x^{NEW}) <= f(x^{CURBEST})$ $x^{CURBEST} = x^{NEW}$ *end if*	*if* $f(x^{NEW}) <= B$ *or* $f(x^{NEW}) <= f(x^{WORST})$ $x^{WORST} = x^{NEW}$ *end if*	*GDGD*

The RC operator which was selected based on 1 - HMCR probability moved or swapped the lecture at j randomly to other timeslots (whether in the same room and timeslot or different room and time slot) that were available and feasible.

In NN and NGD, for each movement of selected neighborhood structure on the MC operator, the quality of the new solution $f(x^{NEW})$ was calculated and compared with the quality of the best solution, $f(x^{CURRBEST})$. If there is an improvement, where

$f(x^{NEW})$ is less or equal to $f(x^{CURRBEST})$, the new solution x^{NEW} is accepted and $x^{CURRBEST}$ is set to the new solution x^{NEW}. The worse solution cannot be accepted. In other hand, in GDN and GDGD, the better solution ($f(x^{NEW})$ whether it is less or equal to $f(x^{CURRBEST})$) is always accepted while the worse solution in which x^{NEW} is more than the best solution $x^{CURRBEST}$, the new solution is accepted if it is less or equal to the value of current water level B.

In these proposed hybridization of HSA with GD algorithms, the water level B did not use any decay rate, instead the value of B was set to the value of the updated best solution in the HM at every MI iteration. In other word, the same water level B was used within N variables iteration.

With the probability of 1-HMCR, the RC operator randomly moved the lecture to any feasible timeslot or swapped the lecture with another lecture located in the same or different rooms or periods. In NN and GDN, the execution of these moves and swap were calculated and compared with the quality of the best solution, $f(x^{CURRBEST})$. If there is an improvement in which $f(x^{NEW})$ is less or equal to $f(x^{CURRBEST})$, the new solution x^{NEW} is accepted and $x^{CURRBEST}$ is set to a new solution x^{NEW}. The worse solution cannot be accepted. In NGD and GDGD, the execution of the RC operator was calculated and compared using the GD acceptance formula. The improved solution ($f(x^{NEW})$ whether it is less or equal to $f(x^{CURRBEST})$), it is always accepted while the worse solution is accepted if it is less or equal to the value of current water level B.

At the end of N variables iteration, the new solution x^{NEW} was updated to HM if the cost of new solution $f(x^{NEW})$ is less or equal to the worst solution in the HM, $f(x^{WORST})$. For HSA and GD hybridization (NGD, GDN, and GDGD), if the new solution x^{NEW} is worse than the worst solution in HM, the new solution is accepted if it less or equal to the value of current water level B. At the beginning of the next MI iteration, the water level B is set to the best solution found so far.

The whole procedure was repeated until the termination criteria (number of MI) were met and the best solution found was returned to the end of this algorithm.

4   Experimental Results

The proposed methods were coded using C++ in Microsoft Visual 2008 under Windows 7 on an Intel Machine with Core TM i7 4770 CPU and a 3.1GHz processor and 3GB RAM. Twenty-one data of instances were categorized as ITC-2007 which were available at CBCTT website (http://tabu.diegm.uniud.it/ctt). The data were used to compare the performance of NN, NGD, GDN, and GDGD. For each proposed algorithm, three HMCR, i.e. 0.2, 0.5, and 0.8 were executed 10 times for each data, by imposing 1000 iterations as the stopping condition.

Table 3 shows the penalties obtained from the best results of 10 run which were obtained by NN, NGD, GDN, and GDGD algorithm with different HMCR setting. The total penalties from Table 3 are highlighted in a bar chart as shown in Figure 2. It is apparent from Figure 2 that the lowest total penalties were obtained from NGD with HMCR 0.5 setting.

Table 3. Results of NN, NGD, GDN and GDGD with different HMCR

HMCR / Data Instances	NN			NGD			GDN			GDGD		
	0.2	0.5	0.8	0.2	0.5	0.8	0.2	0.5	0.8	0.2	0.5	0.8
Comp01	6	5	6	7	5	6	8	7	6	9	7	10
Comp02	151	145	143	116	81	111	144	127	127	129	125	125
Comp03	162	146	151	116	108	103	150	136	147	141	144	139
Comp04	74	73	75	81	62	57	90	88	100	105	107	105
Comp05	479	524	514	381	365	396	441	412	391	361	365	358
Comp06	118	110	113	116	100	81	127	138	152	161	149	157
Comp07	112	102	100	111	73	61	123	124	154	156	161	176
Comp08	87	81	88	91	76	58	100	115	130	129	131	128
Comp09	181	165	164	149	136	142	177	166	180	180	178	185
Comp10	82	82	85	77	52	53	102	111	119	128	132	137
Comp11	0	0	0	0	0	0	0	0	0	1	1	0
Comp12	584	557	584	385	392	454	480	482	412	424	406	416
Comp13	124	120	110	113	103	95	128	139	153	153	152	153
Comp14	108	110	116	94	81	82	120	114	118	122	118	122
Comp15	153	167	158	111	116	124	153	142	137	143	147	148
Comp16	108	89	100	108	81	68	129	129	140	153	147	160
Comp17	142	149	137	140	117	113	162	180	176	174	172	173
Comp18	113	115	115	104	106	117	115	105	100	101	106	104
Comp19	182	154	137	107	104	107	139	129	138	126	135	136
Comp20	107	117	130	90	82	86	136	135	148	143	135	140
Comp21	205	197	206	167	143	169	206	207	205	209	211	214
TOTAL	3278	3208	3232	2664	**2383**	2483	3230	3186	3233	3248	3229	3286

It can be concluded that the hybridization of HSA with GD, with the range of 50% normal acceptance formula (the movement is accepted if the cost is less or equal to the current cost) in MC operator and 50% of the GD acceptance formula in RC operator (NGD) outperformed those obtained by other HSA and GD hybridization algorithms with the lowest total penalties, i.e. 2383. NGD with HMCR 0.8 gave a better performance compared to NGD with HMCR 0.2. These results demonstrated that the use of less than 50% of GD acceptance formula in the RC operator produces better performance. In addition, these results indicated that the MC operator which used different neighborhood structures with normal acceptance formula contributed faster convergence rather than employing the neighborhood structures with the GD acceptance formula. This was verified with the result provided by GDN and GDGD, in which both of them applied GD acceptance formula in the MC operator.

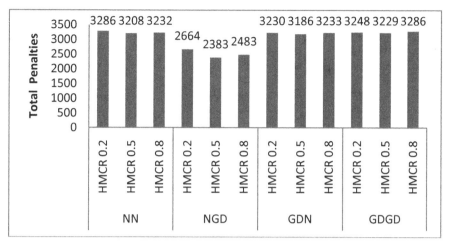

Fig. 2. Total Penalties of HSA and GD Hybridization

4.1   Experiments with Higher Number of Iterations

After further analysis on the performance graph of the NGD algorithm, it was found that the improvements are still being made towards the end of the maximum number of iterations, i.e. 1000. Therefore, the execution of NGD was extended until no further improvement for the last 1000 iterations. This is to study whether much better solutions can be found.

The execution of NGD with higher iteration was applied to all problem instances except for Comp01 and Comp05, as both of the problem instances were already at the state of optimal solution.

Table 4 shows the percentage improvement of NGD with higher iterations compared to NGD with 1000 iterations. The NGD with higher iterations was able to improve the solution for most of the problem instances except Comp05. The NGD with higher iterations was able to improve the solution by more than 10% for eleven problem instances.

4.2   Comparing with Other Approaches in Literature

Table 5 shows the comparison of NGD with higher iterations with other approaches in the literature. The objective here was to show that the hybridization of HSA and GD is able to produce a good quality and feasible solutions for CBCTT problems even though they may not produce the best results. The approaches considered include repair-based heuristic using propositional satisfiability (SAT) by [11], dynamic tabu search by [7], combination of hill climbing (HC), great deluge (GD) and simulated annealing (SA) by [12], great deluge (GD) with kempe chain neighborhood by [13], adaptive tabu search (ATS) by [14], adaptive tabu search (ATS) with numerous combination of neighborhoods by [15], integer programming by [16], threshold accepting by [17], combination of an electromagnetic-like mechanism (EM) and great deluge (GD) by [18], propositional satisfiability (SAT) solvers and optimizers by [19], combination of simulated annealing (SA) and dynamic tabu search (DTS) by [20], and hybridization between simulated annealing and non-accepted solutions memory (SAM) by [21].

Table 4. Improvement Percentage of NGD with Higher Iterations

Problem instances	NGD HMCR 0.5 1000 iterations	NGD HMCR 0.5 Higher iterations		
	Penalty	Penalty	No. of iterations	Improvement Percentage
Comp01	5	5	Optimal	Optimal
Comp02	81	66	2177	18.52
Comp03	108	98	3984	9.26
Comp04	62	43	5259	30.65
Comp05	365	365	441	0.00
Comp06	100	78	5445	22.00
Comp07	73	30	7544	58.90
Comp08	76	50	4295	34.21
Comp09	136	126	2912	7.35
Comp10	52	36	5013	30.77
Comp11	0	0	Optimal	Optimal
Comp12	392	390	1893	0.51
Comp13	103	89	4882	13.59
Comp14	81	69	2309	14.81
Comp15	116	110	3495	5.17
Comp16	81	47	4544	41.98
Comp17	117	98	2456	16.24
Comp18	106	103	2333	2.83
Comp19	104	94	3719	9.62
Comp20	82	68	3611	17.07
Comp21	143	132	1981	7.69
TOTAL	2383	2097		12.00

The dashes (-) sign in Table 5 indicates the problem instances that were not experimented by the authors. The best results were highlighted in each cell. The first comparison resulted against [11] who had applied repair-based heuristic using propositional satisfiability (SAT). The NGD results were better than the repair-based heuristic in all problem instances. The NGD results were better than [7] for three problem instances (Comp02, Comp04, and Comp07), better than [16] for two problem instances (Comp01 and Comp12), better than [19] for four problem instances (Comp03, Comp05, Comp12, and Comp15), and has equal result with [21] for one problem instance (Comp04).

In Table 5, the comparisons were also made with the results obtained from the best-known solution (last column) available from CBCTT website (http://satt.diegm.uniud.

it/ctt). The NGD algorithm obtained the optimal solution for Comp01 and Comp11, while achieved competitive results with the best-known solution for the rest of the problem instances.

Table 5. Comparison of Results Obtained by the NGD (HMCR 0.5 higher iteration) with Other Approaches

Problem Instances	NGD Penalty	[11] Penalty	[7] Penalty	[12] Penalty	[13] Penalty	[14] Penalty	[15] Penalty	[16] Penalty	[17] Penalty	[18] Penalty	[19] Penalty	[20] Penalty	[21] Penalty	Best Known Solution (Until 01/07/2015) Penalty
Comp01	5	9	5	5	5	5	5	13	5	5	5	5	5	5
Comp02	66	103	75	43	60	34	40	43	91	39	24	41	35	24
Comp03	98	101	93	72	81	70	71	76	108	76	111	66	77	64
Comp04	43	55	45	35	39	38	39	38	53	35	35	35	43	35
Comp05	365	370	326	298	321	298	298	314	359	315	1343	301	293	284
Comp06	78	112	62	41	45	47	47	41	79	50	27	43	51	27
Comp07	30	97	38	14	21	19	21	19	36	12	6	18	15	6
Comp08	50	72	50	39	41	43	43	43	63	37	37	39	46	37
Comp09	126	132	119	103	102	99	101	102	128	104	171	96	99	96
Comp10	36	74	27	9	17	16	18	14	49	10	4	15	6	4
Comp11	0	1	0	0	0	0	0	0	0	0	0	0	0	0
Comp12	390	393	358	331	349	320	320	405	389	337	977	320	307	294
Comp13	89	97	77	66	73	65	65	68	91	61	59	64	71	59
Comp14	69	87	59	53	59	52	55	54	81	53	51	53	55	51
Comp15	110	119	87	84	82	69	-	-	-	73	111	66	68	62
Comp16	47	84	47	34	49	38	-	-	-	32	18	28	32	18
Comp17	98	152	86	83	81	80	-	-	-	72	56	71	61	56
Comp18	103	110	71	83	79	67	-	-	-	77	83	69	70	61
Comp19	94	111	74	62	67	59	-	-	-	60	57	60	62	57
Comp20	68	144	54	27	30	35	-	-	-	22	4	29	14	4
Comp21	132	169	117	103	110	105	-	-	-	95	86	89	81	74

5   Conclusion and Future Work

The overall goal of this paper was to investigate the HSA hybridization with great deluge for solving the CBCTT problem. The hybridization produced three proposed algorithms such as hybridization of GD in RC operator of HSA (NGD), hybridization of GD in MC operator of HSA (GDN), and hybridization of GD in MC and RC operator of HSA (GDGD). In addition, each proposed algorithm was executed using three different harmony memory consideration rate (HMCR) such as 0.2, 0.5, and 0.8. The performance of each proposed algorithm with different HMCR was compared to each other based on the lowest total penalties obtained. The NGD with HMCR 0.5 produced the lowest total penalties compared to all other proposed algorithms. A further execution of NGD with HMCR 0.5 using higher number of iterations was carried out to find whether the solution of each problem instances can be improved. The results showed more than 10% of improvement for half of the problem instances. The result of NGD with HMCR 0.5 (with higher number of iterations) was compared to other approaches in the literature which applied the same domain and the best-known solution available in the CBCTT website. The approach produced solutions that were better than one published results of all problem instances, while it was better on certain problem instances on certain published results. Moreover, this approach was able to obtain the optimal penalty cost for two problem instances. In the future, this proposed approach can be applied to real data CBCTT problem.

References

1. Gomes, C.P., Williams, R.: Approximation Algorithms. In: Kendall, G., Burke, E.K. (eds.) Search Methodologies Introductory Tutorials in Optimization and Decision Support Techniques. Springer Science-i-Business Media, LLC (2005)
2. Wolpert, D.H., Macready, W.G.: No free lunch theorems for optimization. IEEE Trans. Evol. Comput. 1(1), 67–82 (1997), doi:10.1109/4235.585893
3. Blum, C., Roli, A.: Metaheuristics in combinatorial optimization:Overview and conceptual comparison. ACM Comput. Surv. 35(3), 268–308 (2003)
4. Al-Betar, M.A., Khader, A.T., Zaman, M.: University Course Timetabling Using a Hybrid Harmony Search Metaheuristic Algorithm. IEEE Trans. on Syst. Man, Cybern. Part C Appl. Rev. 42(5), 664–681 (2012)
5. Jaradat, G.M., Ayob, M.: A Comparison between Hybrid Population-based Approaches for solving Post-Enrolment Course Timetabling Problems. IJCSNS Int. J. Comput. Sci. Netw. Security 11(11), 116 (2011)
6. Geiger, M.: Multi-criteria Curriculum-Based Course Timetabling—A Comparison of a Weighted Sum and a Reference Point Based Approach. In: Ehrgott, M., Fonseca, C., Gandibleux, X., Hao, J.-K., Sevaux, M. (eds.) Evolutionary Multi-Criterion Optimization, vol. LNCS, vol. 5467, pp. 290–304. Springer, Heidelberg (2009)
7. De Cesco, F., Di Gaspero, L., Schaerf, A.: Benchmarking curriculum-based course timetabling: Formulations, data formats, instances, validation, and results. In: Proceedings of the Seventh PATAT Conference (2008).
   http://tabu.diegm.uniud.it/ctt/DDS2008.pdf
8. Geem, Z.W., Kim, J.H., Loganathan, G.V.: A new heuristic optimization algorithm: harmony search. Simulation 76, 60–68 (2001)
9. Dueck, G.: New Optimization Heuristic: The Great Deluge Algorithm and the Recordto-record Travel. J. Comput. Phys. 104, 86–92 (1993)
10. Dreo, J., Siarry, P., Petrowski, A., Taillard, E.: Metaheuristics for Hard Optimization, pp. 153–176. Springer, Heidelberg (2006)
11. Clark, M., Henz, M., Love, B.: QuikFix: A repair-based timetable solver. In: Proceedings of the Seventh PATAT Conference (2008).
    http://www.comp.nus.edu.sg/~henz/publications/ps/
    PATAT2008.pdf
12. Müller, T.: ITC2007 solver description: a hybrid approach. Ann. Oper. Res. 172(1), 429–446 (2009)
13. Shaker, K., Abdullah, S.: Incorporating great deluge approach with kempe chain neighbourhood structure for curriculum-based course timetabling problems. In: 2nd Conference on Data Mining and Optimization, DMO 2009, October 27-28, pp. 149–153 (2009)
14. Lü, Z., Hao, J.-K.: Adaptive Tabu Search for course timetabling. Eur. J. Oper. Res. 200(1), 235–244 (2010)
15. Lü, Z., Hao, J.-K., Glover, F.: Neighborhood analysis: a case study on curriculum-based course timetabling. J. Heuristics, 1–22 (2010)
16. Lach, G., Lübbecke, M.: Curriculum based course timetabling: new solutions to Udine benchmark instances. Ann. Oper. Res., 1–18 (2010)
17. Geiger, M.: Applying the threshold accepting metaheuristic to curriculum based course timetabling. Ann. Oper. Res., 1–14 (2010)
18. Abdullah, S., Turabieh, H., McCollum, B., McMullan, P.: A hybrid metaheuristic approach to the university course timetabling problem (2010)

19. Asín Achá, R., Nieuwenhuis, R.: Curriculum-based course timetabling with SAT and MaxSAT. Ann. Oper. Res., 1–21 (2012)
20. Bellio, R., Di Gaspero, L., Schaerf, A.: Design and statistical analysis of a hybrid local search algorithm for course timetabling. J. Sched. 15(1), 49–61 (2012)
21. Tarawneh, H.Y., Ayob, M., Ahmad, Z.: A Hybrid Simulated Annealing with Solutions Memory for Curriculum-based Course Timetabling Problem. J. Appl. Sci. 13, 262–269 (2013), doi:10.3923/jas.2013.262.269; ISSN 1812-5654

A Parallel Latent Semantic Indexing (LSI) Algorithm for Malay Hadith Translated Document Retrieval

Nurazzah Abd Rahman, Zulaile Mabni, Nasiroh Omar,
Haslizatul Fairuz Mohamed Hanum, and Nik Nur Amirah Tuan Mohamad Rahim

Faculty of Computer and Mathematical Sciences
Universiti Teknologi MARA 40450 Shah Alam, Malaysia
{nurazzah,zulaile,nasiroh,fairuz}@tmsk.uitm.edu.my,
nikamirah@gmail.com

Abstract. Latent Semantic Indexing (LSI) is one of the well-known searching techniques which match queries to documents in information retrieval applications. LSI has been proven to improve the retrieval performance, however, as the size of documents gets larger, current implementations are not fast enough to compute the result on a standard personal computer. In this paper, we proposed a new parallel LSI algorithm on standard personal computers with multi-core processors to improve the performance of retrieving relevant documents. The proposed parallel LSI was designed to automatically run the matrix computation on LSI algorithms as parallel threads using multi-core processors. The Fork-Join technique is applied to execute the parallel programs. We used the Malay Translated Hadith of Shahih Bukhari from Jilid 1 until Jilid 4 as the test collections. The total number of documents used is 2028 of text files. The processing time during the pre-processing phase of the documents for the proposed parallel LSI is measured and compared to the sequential LSI algorithm. Our results show that processing time for pre-processing tasks using our proposed parallel LSI system is faster than sequential system. Thus, our proposed parallel LSI algorithm has improved the searching time as compared to sequential LSI algorithm.

Keywords: Latent Semantic Indexing (LSI), Parallel programming, Fork-Join.

1   Introduction

Searching through a huge collection of Web documents is very difficult and time consuming. Prior to searching, an indexing table must be built to assist for a more convenient and efficient search. Inverted Files, Suffix Array and Signature Files are among the common indexing techniques applied to large documents. However, searching has moved from just looking at simple word matching strategies into content-based search. In order to assist search on similar or synonyms word, Latent Semantic Indexing (LSI) technique offers better approach to extract information from large collections of text documents [1]. LSI can be used to do automatic indexing and information retrieval by mapping documents as well as terms to a

© Springer Science+Business Media Singapore 2015
M.W. Berry et al. (Eds.): SCDS 2015, CCIS 545, pp. 154–163, 2015.
DOI: 10.1007/978-981-287-936-3_15

representation in the Latent Semantic space. In other words, LSI can overcome the exact term-matching problem by automatically discovering latent relationship in the document collection and by retrieval based in the higher level semantic structure rather than just the surface level word choice [2]. LSI uses method from linear algebra called Singular Value Decomposition (SVD) that requires matrix computation on term-document matrices. The SVD method is used to discover the important associative relationship between term and term, between term and document and between document and document. The core of the SVD algorithm requires an eigen decomposition of the matrix, which has been a computational problem for many decades [3]. The SVD computation has taken a lot of time which makes it a prime candidate for decreasing processing times [3]. With hundreds on documents to be computed, hundreds number of matrices with big dimensions need to be handled at appropriate speed efficiently.

Information retrieval systems often have to deal with very large amounts of data. Computer must be able to process many gigabytes or even terabytes of text, and to build and maintain an index for millions of documents. A single computer simply does not have the computational power or the storage capabilities required for indexing even a small fraction of the World Wide Web. As the data volume and query processing loads increase, mechanisms are needed to improve the performance of information retrieval. The idea is to partition large document collections, as well as their index structures, across computers. This not only allows for larger storage capacities, but also permits searches to be executed in parallel [4]. Therefore, in this paper, we proposed a parallel LSI which was designed to automatically run the matrix computation on LSI algorithms as parallel threads using multi-core processor. The Fork-Join technique is applied to execute the parallel programs [5]. Our parallel algorithm was applied on the matrices to improve the computation and performance on information retrieval. The paper is organized as follows. Section 2 reviews some of the related research on LSI. In Section 3, the proposed framework and methods for the parallel LSI are presented. Section 4 discusses on the result of our experiments and section 5 concludes the paper.

2   Related Works

The most popular indexing methods in the text retrieval are inverted files, suffix arrays and signature files. The other indexing technique which is widely used is Latent Semantic indexing (LSI). LSI is an effective automated method for determining if a document is relevant to a reader based on a few words [6].

Kowalski [7] identifies that indexing by term usually uses thesauri or dictionary or any other expansion technique to expand a query to find ways the same thing has been represented. On the other hand, LSI does not require the same technique as uses by indexing by term. LSI is a statistical technique that derives a statistical correlation between all terms and documents in corpus, in an attempt to overcome the problem inherent in lexical matching [8]. LSI can be described as a mathematical technique that employed Singular Value Decomposition (SVD) to create relationship between terms and terms, terms and documents, and documents and documents in reduced

Latent Semantic Space. SVD is a technique in linear algebra that discovers the important associative relationship between elements in the matrix. Since LSI is based on the relationship of vectors of terms and documents, and the associations are derived from numerical analysis of existing texts or terms or documents, therefore no external dictionaries, thesauri or knowledge bases are necessary [9].

LSI is an important step to the document indexing process and provides a good solution for a wide range of information retrieval [10]. To date, LSI is routinely applied to collections of millions to tens millions of documents since LSI works better than a pure keyword search. LSI use the same process as keyword searches, but for LSI, the keyword search is applied to a new set of words counts for each document known as the term-document matrix, that are derived from the original documents [6].

There are some related researches on using LSI for information retrieval. Muhamad Taufik et al. [11], has experimented on LSI technique which was applied to Malay language retrieval environment. They worked on the actual Malay Quranic collection and the actual English Quranic collection that contain of 6236 documents in each language. They concluded that LSI work well for Malay-English cross language and monolingual information retrieval.

Another research has worked on LSI for Malay text retrieval [12]. In this research, the Singular Value Decomposition (SVD) method was applied to model documents and queries as vectors in reduced space. The components of the vector were determined by the term weighting scheme, a function of the frequencies of the terms in the document or query.

Cavanagh, et al. [3] has implemented parallel Latent Semantic Analysis (LSA) using a Graphics Processing Unit (GPU). The performance of the parallel LSA on the GPU is compared to traditional LSA implementation on CPU. The results showed that, for large matrices that have dimensions divisible by 16, the GPU algorithm ran five to six faster that the CPU version.

A recent research on LSI was done by Sadjirin and Rahman[13] which applied LSI to a total of 210 Malay language documents. For the test collections, 95 documents are taken from a collection of Sahih Muslim's hadith and 115 documents from a collection of Sahih Bukhari's hadith. Their experiment results show that for the Malay document retrieval, LSI performed 40% better compared to Inverted Index Files technique. However, the processing time for LSI is higher as compared to that of Inverted File technique. Thus, LSI has high computational cost for the matrix decomposition.

3   The Proposed Framework and Techniques

This section describes the framework for the prototype of the system. Fig. 1 illustrates the framework of Latent Semantic Indexing for Retrieving Malay Hadith Translated Document. The first component in this framework is the test collections named Collection of Translated Hadith (Shahih Bukhari). Next component is Latent Semantic Space which is a vector space model representing the location or the vector of terms and documents in LSI. The Term-by-Document Matrix is a matrix representation that consists of term, occurrences of term in the document as well as the documents or articles itself. In this matrix, columns represent the document, and rows represent the terms.

Query processing is the task that processes query words which is formulated by the user. During query processing, the query term is stemmed and reduced to the root word. The query is measured against the terms vector and documents vector in the Latent Semantic space based on cosine similarity [1, -1]. The result returned to the user is depending on the ranking instead of exact term-matching strategies. If the document ranking is 1, it means that the document retrieved is exactly similar in the context, while -1 indicates that there is no similarity at all. Hadith document files will be retrieved and displayed if the queries have similarities against terms documents.

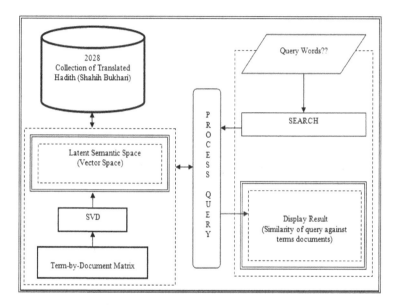

Fig. 1. Latent Semantic Indexing Framework

3.1   Pre-processing

For the proposed new Parallel Latent Semantic Indexing technique, the pre-processing tasks used the Fork-Join technique to break down the problem to simple tasks until these become simple enough to be solved directly and make the processing faster. Several pre-processing tasks that need to be accomplished are as follows:

Extract Terms (List of Terms) from Document Files in the Test Collection.
During the process of terms extraction, the Malay stopword and special characters are removed from document's contents and then stemmed into its root word by removing its prefixes and suffixes. The parallel process of extracting term documents in test collection was done by using Fork-Join technique. Firstly, the major process is divided into several parts or tasks. This part is processed on a separate CPU. Each part will start the process to read the document's name from test collection, extract all the contents, and converting all the string into lower case. Then, the removal of stopword and stemming procedures is performed. Next, each word is compared to the other

words. The same or identical word is grouped together. Before proceeding to the next step of Fork-Join technique, each part is processed until all the files have been retrieved and all the contents from document files have been extracted. Finally, all parts are joined after all documents from test collection have been retrieved. Lastly, the list of terms are sorted and saved into a file named 'Term_List'.

Create Term-by-Document Matrix.
Term-by-document matrix is created by comparing a term that has been created against the document from test collection. During the process of creating term-by document matrix, the Malay stopword and special characters are removed from document's contents and then stemmed into its root word by removing its prefixes and suffixes. Term-by-document matrix file is the most important and critical part in this research because it will be used in Singular Value Decomposition (SVD) computation to derive the Latent Semantic space. Firstly, the system will read the 'Term_List' file and documents from test collection. The occurrences of term from 'Term_List' file will be compared and counted against the subsets (group documents) from test collection. If the term does not occur, the system will write 0, otherwise it will write the number the occurrences. The list of occurrences will be in matrix form, where the rows represent the term and the columns represent the documents. The system will repeat the same process until the entire documents from test collection has been retrieved. If there are no more documents to be processed, the term-by-document will be saved in a file named 'Term_Document_Matrixs' file.

Create the LSI Space with the Specific k Value.
The process of creating or calculating the LSI/LSA space is the process that takes the longest time in sequential process. In our proposed framework, Fork-Join technique is applied to this process in order to make the searching part faster. For this process, we need to choose the right value of k that will be use in creating the LSI space. Very low values of k means user may lose some of the relevant document. But very high values of k may not change the result much from the simple vector search. Firstly, parallel process started by setting the k value, retrieving all the terms in the 'Term_List' file and all the matrices in the 'Term_Document_Matrix' file, and reading all the document files. Next, the process is divided into several parts or tasks. Each part is processed on a separate CPU. The relationship of term-by-term and term-by-document representation in LSI is derived by using SVD method. The SVD method created the LSI Space in parallel threads. Lastly, the matrix values are saved to files named 'mRowVectors' and 'mColumnVectors'.

3.2   Data Collection

Hadith Documents.
In this research, we used 2028 documents, which are derived from Malay translated hadith from Shahih Bukhari collections. Table 1 below shows several examples of hadith documents in the test collections.

The name of each Malay document is uniquely identified according to its naming convention. For example, file name 'H0001MJ1' indicates that the letter 'H' means the file is a Malay hadith document, the next four digit ('0001') means number of document number, 'M' means the hadith is Muslim's hadith, while 'J' means 'Jilid' or volume, and the last digit ('1')indicates the number of volumes of the Malay hadith document.

Table 1. Example of Hadith Document

Document Name	Example of Document's Content
H0927MJ2	927. Dari Abu Sa'id Al Khudri r.a., dari Nabi saw., sabdanya : "Tidak wajib dizakatkan bahan makanan pokok yang kurang dari lima wasq tidak pula binatang ternak yang kurang dari lima ekor; dan emas perak yang kurang dari lima uqiah."
H2384MJ4	2384. Dari Ibnu 'Umar r.a., dari Nabi saw. sabdanya: "Perumpamaan orang munafik, seperti seekor kambing betina yang bingung mengikuti dua ekor kambing jantan.Sekali dia hendak mengikuti yang ini, kali yang lain hendak mengikuti yang itu."

Sample of Query.
Query word is a string of word that characterizes the information searched by the user. The query formulated by the user is in a short natural language query or statement. There are 315 queries listed with respect to each category of the hadith document such as 'Iman', 'Zakat' and 'Solat' [15]. In this research, 10 queries are selected to be used as a sample for retrieval of relevant document in test collection as shown in Table 2.

Relevant Judgment.
Relevant judgment is used to determine the degree of relevant document retrieved by the user but different user will have different judgment on the documents because it depends on their satisfaction. User may retrieve few documents in response to their queries. Nevertheless, not all documents retrieved are relevant to user request.

In information retrieval experiments, the relevance of documents that are retrieved in response to each query is assessed for the effectiveness. The list of relevant judgment is one of the elements of the hadith test collection. List of relevant judgment is provided by hadith Shahih Bukhari's book.

Table 2. List of Sample Query

Query	Query Words
1	Tuntutlah ilmu hingga ke liang lahad
2	Adab-adab berkaitan makan dan minum
3	Hormati kedua ibu bapa
4	Hukum hudud
5	Hukum bernazar
6	Penghijrahan ke Madinah
7	Apakah hukum berhias bagi kaum wanita
8	Bagaimana cara solat Jenazah
9	Pembahagian harta mengikut
10	Balasan di hari kiamat

Table 3 below shows some of the relevant judgment for 2 queries among the 12 queries tested.

Table 3. Sample of relevant judgment

	Query Words	Relevant Judgement
1	Tuntutlah ilmu hingga ke liang lahad	H0005BJ1, H0049BJ1, H0056BJ1
2	Adab-adab berkaitan makan dan minum	H0111BJ1, H0848BJ2, H0945BJ2

4   Result and Discussion

This section discusses about our experimental results.

Table 4. Time taken during terms extraction for sequential and parallel system

Number of Files to be processed	Extract Term for Sequential system (second)		Extract Term for Parallel system (second)	
	M2	M4	M2	M4
200	0:393	0:294	0:350	0:172
400	0:513	0:323	0:506	0:236
600	0:905	0:504	0:647	0:302
800	1:125	0:629	0:724	0:359
1000	1:592	1:039	0:880	0:406

The first experiment compares the time taken during terms extraction for sequential and parallel LSI system. Two machines were used, one machine contains two processors (M2) and another one contains four processors (M4). Table 4 below illustrates the time taken during extract term list in pre-processing and tested it using two difference machines for both systems.

The results in Table 4 indicate that the time taken to extract term list for the parallel system is faster than sequential system using machine with two processors (M2) as well as four processors (M4).

Fig. 2 below shows that the time taken to extract term list for the parallel system is faster than sequential system using machine with two processors (M2).

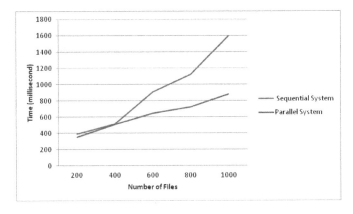

Fig. 2. Graph representing the time taken to extract term list for both systems using two processors (millisecond)

Meanwhile, Fig. 3 below shows that the time taken to extract term list for the parallel system is faster than sequential system using machine with four processors (M4).

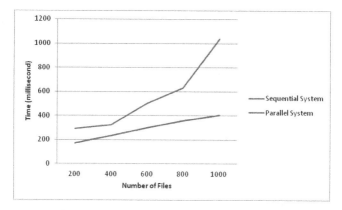

Fig. 3. Graph representing the time taken to extract term list for both systems using four processors (millisecond)

The second experiment compares the time taken of creating LSI space for both systems between two difference machines, one machine contains two processors (M2) and another one contains four processors (M4). Table 5 below illustrates the time taken for creating LSI space in pre-processing and tested it in two difference machines.

Table 5. Time taken for creating LSI space for sequential and parallel system

Number of Files to be processed	Create LSI Space for Sequential System (second)		Create LSI Space for Parallel System (second)	
	M2	M4	M2	M4
200	60:172	47:420	13:761	9:718
400	190:088	127:110	35:856	25:563
600	379:491	209:967	53:259	41:864
800	542:197	318:397	78:540	62:510
1000	771:306	427:852	101:875	82:338

Table 5 shows that for a total of 600 files processed, the time taken to create LSI space for sequential system and parallel system using machine with two processors are 379:491 and 53:259 seconds respectively. On the other hand, for the same number of 600 files processed, the time taken to create LSI space for sequential system and parallel system using machine with four processors are 209:967 and 41:864 seconds respectively. Thus, the overall results indicate that the time taken to create LSI space for parallel system is faster than sequential system using machine with two processors as well as four processors.

5   Conclusion

In this paper, we have proposed a new parallel LSI algorithm on standard personal computers with multi-core processors to improve the performance of retrieving relevant documents. The proposed parallel LSI was designed to automatically run the matrix computation on LSI algorithms as parallel threads using multi-core processor. The Fork-Join technique is applied to execute the parallel programs. We have used the Malay Translated Hadith Shahih Bukhari from Jilid 1 until Jilid 4 as the test collections. Our results show that the processing time for preprocessing tasks using our proposed parallel system is faster than sequential system. Thus, our proposed parallel LSI algorithm has improved the searching time as compared to sequential LSI algorithm.

Acknowledgments. We wish to acknowledge the support of Malaysian Ministry of Higher Education (MOHE) under the Fundamental Research Grant Scheme (FRGS) No: FRGS/2/2010/SG/UITM/02/12.

References

1. Dumais, S., Furnas, G., Lanerwester, T., Harshmandauer, R., Deerwester, S.: R. Harshman, R.: Using Latent Semantic Analyses to Improve Access to Textual Information. In: Proceedings of the SIGCHI Conference on Human factors in Computing Systems, Washington, D.C., United States, pp. 281–285 (May 1988)
2. Foltz, W.P., Dumais, T.S.: Personalized Information Delivery: An Analysis of Filtering Methods. Communications of the ACM 35(12), 51–60 (1992)
3. Cavanagh, J.M., Potok, T.E., Chiu, X.: Parallel Latent Semantic Analysis using Graphics Processing Unit. In: Proceedings of the 11th Annual Conference Companion on Genetic and Evolutionary Computation Conference, pp. 2505–2510. ACM, New York (2009)
4. Tomasic, A., Garcia-Molina, H.: Issues in Parallel Information Retrieval. Stanford University (1994)
5. http://www.cs.cmu.edu/~tomasic/doc/1994/
   TomasicGarciaDEB1994.pdf
6. Tsagklis, I.: Java Fork/Join for Parallel Programming (2011),
   http://www.javacodegeeks.com/2011/02/
   java-forkjoin-parallel-programming.html
7. Story, R.E.: An Explanation of the Effectiveness of Latent Semantic Indexing By Means of a Bayesian Regression Model. Information Processing and Management 32, 329–344 (1996)
8. Kowalski, G.: Cataloging and Indexing. In: Information Storage and Retrieval Systems, vol. 8, pp. 51–69. Springer, USA (2008)
9. Wahlen, N.: A Comparison of Different Parallel Programming Models for Multicore Processors. Bachelor of Science Thesis. KTH Information and Communication Technology, Stockholm, Sweden (2010)
10. Kim, W., Khudanpur, S.: Lexical Triggers and Latent Semantic Analysis For Cross-Lingual Language Model Adaptation. ACM Transactions on Asian Language Information Processing (TALIP), 94–112 (2004)

11. Kontostathis, A., Pottenger, W.M.: A framework for Understanding Latent Semantic Indexing (LSI) Performance. Information Processing and Management 42, 56–73 (2006)
12. Abdullah, M.T., Ahmad, F., Mahmod, R., Sembok, T.M.T.: Application of Latent Semantic Indexing on Malay-English Cross Language Information Retrieval. In: Sembok, T.M.T., Zaman, H.B., Chen, H., Urs, S.R., Myaeng, S.-H. (eds.) ICADL 2003. LNCS, vol. 2911, pp. 663–665. Springer, Heidelberg (2003)
13. Ab Samat, N., Azmi Murad, M.A., Abdullah, M.T., Atan, R.: Term Weighting Schemes Experiment Based on SVD for Malay Text Retrieval. IJCSNS International Journal of Computer Science and Network Security 8, 357–361 (2008)
14. Sadjirin, R., Rahman, N.: Efficient Retrieval of Malay Language Documents using Latent Semantic Indexing. In: International Symposium Information Technology (ITSim), vol. 3, pp. 1410–1415 (2010)
15. Deerwester, S., Dumais, T.S., Furnas, W.G., Landauer, K.T.: Indexing by Latent Semantic Analysis. Journal of the American Society of Information Science 41(6), 391–407 (1990)
16. Kamarul, A.M.: To Evaluate the Effectiveness and Efficiency of Stemming Algorithm and Biagram Method in Retrieving Hadith Documents. BSc. Thesis, Universiti Teknologi MARA (2002)

Short Term Traffic Forecasting Based on Hybrid of Firefly Algorithm and Least Squares Support Vector Machine

Yuhanis Yusof, Farzana Kabir Ahmad, Siti Sakira Kamaruddin,
Mohd Hasbullah Omar, and Athraa Jasim Mohamed

School of Computing, College of Arts and Sciences, Universiti Utara Malaysia,
06010 Sintok, Kedah, Malaysia
yuhanis@uum.edu.my

Abstract. The goal of an active traffic management is to manage congestion based on current and predicted traffic conditions. This can be achieved by utilizing traffic historical data to forecast the traffic flow which later supports travellers for a better journey planning. In this study, a new method that integrates Firefly algorithm (FA) with Least Squares Support Vector Machine (LSSVM) is proposed for short term traffic speed forecasting, which is later termed as FA-LSSVM. In particular, the Firefly algorithm which has the advantage in global search is used to optimize the hyper-parameters of LSSVM for efficient data training. Experimental result indicates that the proposed FA-LSSVM generates lower error rate and a higher accuracy compared to a non-optimized LSSVM. Such a scenario indicates that FA-LSSVM would be a competitor method in the area of time series forecasting.

Keywords: short term forecasting, Least Squares Support Vector Machine, Firefly algorithm, traffic management system.

1   Introduction

Advance traveler information systems (ATIS) can facilitate travellers in making travelling decisions as it provides information on the traffic condition. Nevertheless, generating accurate and timely information to be provided to the travellers is a challenging task. This can be done using three types of information; historical information that is based on archived data, real time information that is based on current situation obtained from systems, and predicted information that can be obtained by developing a traffic forecasting using real time or historical information [1].

The traffic forecasting utilizes historic data to predict or estimate the future event or traffic trends that later help travellers in decision making. Generally, forecasting is divided into two types; short term forecasting and long term forecasting [2, 3, 4]. Long term forecasting provides information on yearly basis of the average daily traffic forecasting. On the other hand, the short term forecasting predicts the variability of traffic flow in short periods (sometimes within minutes) using historical or real time

© Springer Science+Business Media Singapore 2015
M.W. Berry et al. (Eds.): SCDS 2015, CCIS 545, pp. 164–173, 2015.
DOI: 10.1007/978-981-287-936-3_16

traffic data that are collected from roadway sensors [2, 3]. There are three types of factors to be forecast in short term traffic prediction; flow, speed and travel time, but the last two has received more attention. This study focuses on developing short term traffic speed forecasting using historical speed data collected from the road sensors installed on the highway. This study is realized by proposing a hybrid model of Least Squares Support Vector Machine and Firefly Algorithm (FA-LSSVM). The structure of this paper is as follows: section 2 provides related work in forecasting traffic speed a brief description of LS-SVM while section 3 explains the proposed FA-LSSVM. Section 4 describes the undertaken experiments while the results is included in section 5. Finally, conclusion of the study is presented in section 6.

2   Related Work

In literature [2, 3, 4], short term traffic forecasting can be developed using one of the following approaches; statistical, Artificial Intelligent or machine learning. Prediction utilizing statistical approach can be classified into two; univariate and multivariate. The univariate strategy predicts the output ,y, from a trend where it includes the utilization of filtering techniques such as Kalman filtering [5, 6] and Exponential filtering [7, 8], Autoregressive integrated moving average (ARIMA) [9], k nearest neighbour [4], and Bayesian statistical probabilities model [1]. On the other hand, multivariate statistical approach forecasts the y from a trend and other variables. For example, Portugais and Khanal (2014) proposed a speed forecast model that is based on dynamic linear models and Bayesian inference. They used Kalman recursion to forecast the traffic speed in a dynamic state-space model, and integrate speed collected from radar based sensors and validates with the data gathered from Bluetooth traffic.

Despite of the statistical approach obtaining good accuracy, it does not have the ability to address the nonlinear characteristics of short term traffic flow [2]. Thus, to overcome this problem, Artificial Intelligence approach was employed. An example of the approach was the adaption of Neural Network (NN) model. NN was proven successful in many applications such as classification, image processing [10] and also short term traffic forecasting [1]. A neural network model that proposed the hybrid of exponential smoothing and Levenberg-Marquardt (Lm) algorithm for short term traffic forecasting on the Mitchell freeway in Western Australia [7] was a successful example. Even though, NN produces high accuracy, it requires some parameter tuning and the structure is very complicated. In addition, it suffers from local minima during the training process [9]. To overcome this problem, some researchers propose the hybrid of NN and Swarm Intelligence algorithms to enhance the generalization capability of the network and identify the optimal weight and parameters. An example of such work includes the Wavelet Network Model (WNM) [11].

On the other hand, machine learning approach allows computers to learn without being explicitly programmed [12]. One example of machine learning is the Support Vector Machine (SVM) [13] that introduces statistical theory as a large margin algorithm. SVM operates by separating the search space by maximum margin hyper plane that leads to splitting the training samples into two classes. It has two significant parameters; the regularization parameters and Kernel parameters [14]. The aim of SVM is to select appropriate parameters to minimize the generalization error caused by the result of prediction model. SVM has ability to handle higher dimensional data and analyse them. The experimental studies made evidence that SVM is efficient in many data mining fields such as regression and pattern recognition [15]. However, SVM is difficult to be implemented and consumes large time because of the computational process that is based on quadratic programming.

Such a drawback in SVM has led to the Least Squares Support Vector Machine (LS-SVM), a modified version of SVM that has similar advantage of SVM. The LS-SVM applies equality constraints instead of inequality constraints and employs linear programming instead of quadratic programming [16]. In this study, the optimization of LS-SVM parameters is achieved by employing one of the Swarm Intelligence algorithms, i.e Firefly algorithm.

2.1   Least Squares Support Vector Machine

In LS-SVM, it is assumed that there is a training set of N points $\{x_j, y_j\}$, where, x_j refers to input values and the output values are y_j. The estimation function of LS-SVM for nonlinear regression is as in equation 1 [17].

$$y(x) = w^T \varphi(x^j) + b + e_j \tag{1}$$

Where, w refers to the weight vector, $\varphi(x_j)$ is nonlinear function, b refers to bias term and e_j refers to the error between actual and predicted output. The two parameters w and b can be achieved by optimization problem as defined in the following equations [17].

$$\min_{w,b,e} J(w,e) = \frac{1}{2} w^T w + \gamma \frac{1}{2} \sum_{j=1}^{N} e_j^2 \tag{2}$$

Subject to equality constraints

$$y_j = w^T \varphi(x^j) + b + e_j , j = 1, 2, \dots, N$$

Where, e_j is the error variable, J(w, e) refers to loss function, and γ is the adjustable constant. The Lagrangian multiplier function is applied to equation 2 yields.

$$L(w,b,e,\alpha) = J(w,e) - \sum_{j=1}^{N} \alpha_j \{w^T \varphi(x^j) + b + e_j - y^j\} \tag{3}$$

Where, αj refers to Lagrangian multiplier, γ refers to regularization parameters. The conditions for optimality upper function of this problem can be generated through all derivatives set to equal zero, which is formulated as in the following.

$$\frac{\partial L}{\partial w} = 0 \rightarrow w = \sum_{j=1}^{N} \alpha_j \, \varphi(x^j) \tag{4}$$

$$\frac{\partial L}{\partial b} = 0 \rightarrow \sum_{j=1}^{N} \alpha_j = 0$$

$$\frac{\partial L}{\partial e_j} = 0 \rightarrow \alpha_j = \gamma e_j$$

$$\frac{\partial L}{\partial \alpha_j} = 0 \rightarrow w^T \varphi(x^j) + b + e_j - y_j = 0$$

Where, j=1, 2, ..., N. After ignoring w and e_j, the Karush-Kuhn-Tucker (KKT) conditions for optimality transform into a set of linear equations as shown in equation 5.

$$\begin{bmatrix} \alpha \\ b \end{bmatrix} \begin{bmatrix} 0 & y^T \\ y & \Omega + {}^{I}\!/_{\gamma} \end{bmatrix} = \begin{bmatrix} 0 \\ y \end{bmatrix} \tag{5}$$

The LS-SVM model for function estimation becomes.

$$y(x) = \sum_{j=1}^{N} \alpha_j \, K(x, x_j) + b \tag{6}$$

Where, α_j, b are the solutions for linear system in equation (5). $K(x, x_j)$ is the kernel function, where in this study, we use the Radial Basis Function (RBF) kernel as shown in equation 7.

$$K(x, x_j) = e^{\frac{\|x - x_j\|^2}{2\sigma^2}} \tag{7}$$

In RBF, σ is a tuning parameter where such value and the value of γ regularization parameter in equation 2 need to be optimized in order to minimize the generalization error. This study proposes the optimization of the two parameters using Firefly algorithm.

2.2   Firefly Algorithm

Firefly algorithm (FA) is swarm based algorithm that was developed by Xin-Shin Yang (2008). It has the ability to identify global optimal solution efficiently. FA has two important factors; the light intensity and the attractiveness between fireflies. The light intensity of a firefly is related with the objective function f(x). The objective function can be a maximization or minimization depending on the problem.

The attractiveness, β, between fireflies is related with light intensity and changes based on the distance between two fireflies. The flow process of Firefly algorithm can be seen in [18] while the proposed FA-LSSVM is shown below.

```
Input:
Generate initial population of firefly.
Generate initial solution randomly xi (i=1, 2,.., n),
where, xᵢ represents the hyper-parameters of LS-SVM ( ,
).
Determine initial position (X, Y) of each firefly, which
represents the solution.
Process:
Initialize the LS-SVM model using generated solution.
Train the LS-SVM model.
Evaluate the LS-SVM model using Mean Average Percent Er-
    ror (MAPE) using
```

$$MAPE = \frac{1}{N}\left[\sum_{i=1}^{N}\left|\frac{y_i - p_i}{y_i}\right|\right]$$

```
Light Intensity, I, at xᵢ is determine by objective func-
tion, f(xᵢ):
```

$$f(x_i) = 1/(1 + MAPEi)$$

```
Define light absorption coefficient  =1.
While (t < max generation)
For i=1 to N (N is the number of fireflies)
For j=1 to N
```
$$\text{If } (I_i < I_j) \ \{\beta = \beta_0 \exp^{(-Yr_{ij}^2)}$$
$$X^i = X^i + \beta_0 \exp^{(-Yr_{ij}^2)} * (X^j - X^i) + \alpha\varepsilon_i$$
$$Y^i = Y^i + \beta_0 \exp^{(-Yr_{ij}^2)} * (Y^j - Y^i) + \alpha\varepsilon_i$$

```
Train the LS-SVM model with new solution.
Evaluate new solutions and update light intensity.
End For i
End For j
Rank the fireflies and find the current global best, g*.
End While

Output:
Firefly that has the brightest light intensity.
```

3   Proposed Hybrid Firefly Algorithm with Least Squares Support Vector Machine (FA-LSSVM)

In utilizing Firefly algorithm (FA) to optimize the hyper-parameters of LS-SVM (σ, γ), each pair value (σ, γ) represents a possible solution in the search space that has limited boundaries of the parameters. These solutions are evaluated using Mean Average Percent Error (MAPE) as the fitness function, and the values are assigned to each firefly so that it represents the firefly initial light. The fireflies compete between each other where firefly with a brighter light will attract the less bright ones. This process will lead to position change of each firefly that can generate a new solution. After a maximum number of iteration has been completed, the optimal value of parameters is obtained. Hyper parameter values that generate the lowest MAPE represent the best solution (i.e. solution that produces the highest prediction accuracy).

4   Experiments

4.1   Data Description

In this study, real datasets obtained from sensors located in the highway are utilized. The samples covered are from 2-Jan-2014 until 4-Jan-2014. Prior to utilize the data, the dataset was divided into three sub-datasets, 70% for training, 15% for validation and 15% for testing. Table I shows the descriptive statistics of our dataset which includes 3000 cases and 5 variables. The variables are SensorID, Time, DayStatus, Date and Speed.

Table 1. Statistical Description of Data

Variables	N	Minimum	Maximum	Mean	Std. Deviation
SensorID	3000	14901	14902	-	-
Time	3000	0:03:01.000	23:59:35.000	-	-
DayStatus	3000	1	3	-	-
Date	3000	02-Jan-2014	04-Jan-2014	-	-
Speed	3000	3	112	70.79	18.587

For *Sensor ID*, there are only two values; sensor1 that is represented as 14901 and sensor2 known as 14902. The *Time* variable represents the time at which the speed of traffic is captured. For the *Date* variable, it represents the period of when the speed was collected. Further information on the date was also utilized which is the status of the day (*DayStatus*). This includes whether the day is categorized as a work day, weekend or public holiday. The other variable used is the *Speed* that indicates the speed of traffic at a specific time.

4.2   Data Normalization

In order to simplify the training task, data normalization was performed using min-max normalization [10] and its equation is as in equation 8.

$$\bar{x} = {x - x_{min}}/{x_{max} - x_{min}} \qquad (8)$$

Where, $\bar{x}$ is the normalized data, x is the original data, x_{min} refers to minimum value in dataset while x_{max} refers to maximum value.

4.3   Criteria Measurement

The proposed traffic forecasting was later evaluated based on two measures; Mean Average Percent Error (MAPE) as shown in equation 9 and Root Mean Square Error (RMSE) as shown in equation 10 [16, 19].

$$MAPE = \frac{1}{N}\left[\sum_{i=1}^{N}\left|\frac{y_i - p_i}{y_i}\right|\right] \qquad (9)$$

$$RMSE = \sqrt[2]{\frac{\sum_{i=1}^{N}(y_i - p_i)^2}{y_i}} \qquad (10)$$

5   Results

Implementation of the proposed FA-LSSVM is conducted using LSSVMlab toolbox [20]. The parameters setting of FA-LSSVM are tuned as follows; the number of fire-flies is 100, absorption coefficient, γ, is set to 1, and initial attractiveness, β, is also given as 1. As can be seen in Table II, the two optimal parameters; tuning parameter, σ , and regularization parameter , γ,   that was obtained using the proposed FA-LSSVM is 0.054898244 and 0.21578692. With such combination, FA-LSSVM gene-rates the lowest MAPE compared to the one produced by LSSVM, where in FA-LSSVM it is 9.6197 while 14.6400 in LSSVM. Hence, the accuracy of FA-LSSVM is 90.3803 while LSSVM is 85.3600. Furthermore, the RMSE in FA-LSSVM is only 0.0778 while LSSVM produces 0.1057. A sample of visual result for the predicted speed (every 20 and 10 minutes) is illustrated in Figure 1 and 2. It can be observed that the prediction value produced by the proposed FA-LSSVM is closer to the actual value as compared to the ones produced by LSSVM.

Table 2. Results

	FA-LSSVM	**LSSVM**
Υ	0.054898244	0.06837
σ^2	0.21578692	0.9879
MAPE (%)	9.6197	14.6400
RMSE (%)	0.0778	0.1057
Accuracy	90.3803	85.3600

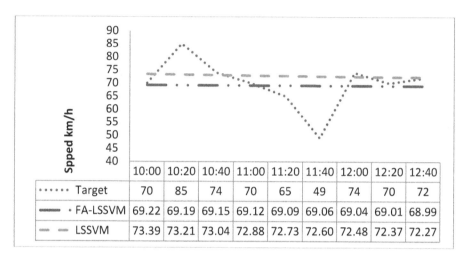

Spped km/h	10:00	10:20	10:40	11:00	11:20	11:40	12:00	12:20	12:40
······ Target	70	85	74	70	65	49	74	70	72
▬ · FA-LSSVM	69.22	69.19	69.15	69.12	69.09	69.06	69.04	69.01	68.99
▭ ▭ LSSVM	73.39	73.21	73.04	72.88	72.73	72.60	72.48	72.37	72.27

Fig. 1. Predicted Traffic Speed for Every 20 minutes

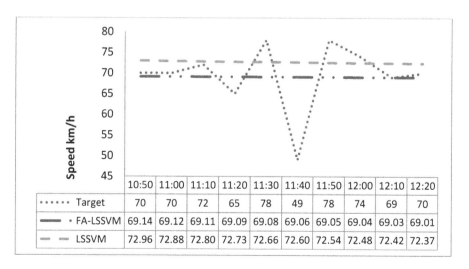

Speed km/h	10:50	11:00	11:10	11:20	11:30	11:40	11:50	12:00	12:10	12:20
······ Target	70	70	72	65	78	49	78	74	69	70
▬ · FA-LSSVM	69.14	69.12	69.11	69.09	69.08	69.06	69.05	69.04	69.03	69.01
▭ ▭ LSSVM	72.96	72.88	72.80	72.73	72.66	72.60	72.54	72.48	72.42	72.37

Fig. 2. Predicted Traffic Speed for Every 10 minutes

6   Conclusion

This study presents a forecasting algorithm to be used in developing an Advance Traveller Information System. The proposed hybridization model between Firefly algorithm and LSSVM, termed as FA-LSSVM, has proven to be a useful forecasting algorithm in predicting the traffic speed for short term period. The optimization of LSSVM hyper-parameters undertaken using Firefly algorithm produces a higher prediction accuracy compared to using LSSVM as an individual forecasting tool. With this, it is hoped that the proposed FA-LSSVM can contribute in developing a more accurate traveler information system that later facilitate users in planning their road trip.

Acknowledgement. The research for this paper is financially supported by the Ministry of Education under the grant s/o 13002. In addition, authors would like to thank Kreatif Apps Sdn. Bhd. and Universiti Utara Malaysia for the collaboration and support given in completing the research.

References

1. Zheng, W., Lee, D., Shi, Q.: Short-term freeway traffic flow prediction: Bayesian combined Neural Network approach. Journal of Transportation Engineering 132(2) (2006)
2. Chan, K.Y., Dillon, T.S., Chang, E.: An Intelligent Particle Swarm Optimization for Short-Term Traffic Flow Forecasting Using on-Road Sensor Systems. IEEE Transactions on Industrial Electronics 60(10), 4714–4725 (2013)
3. Chan, K.Y., Dillon, T.S., Chang, E., Singh, J.: Prediction of Short-Term Traffic Variables Using Intelligent Swarm-Based Neural Networks. IEEE Transactions on Control Systems Technology 21(1), 263–274 (2013)
4. Lin, L., Li, Y., Sadek, A.: A k Nearest Neighbor based Local Linear Wavelet Neural Network Model for On-line Short-term Traffic Volume Prediction. Procedia - Social and Behavioral Sciences 96, 2066–2077 (2013)
5. Meinholda, R.J., Singpurwalla, N.D.: Robustification of Kalman Filter Models. Journal of the American Statistical Association 84(406), 479–486 (1989)
6. Portugais, B., Khanal, M.: Adaptive Traffic Speed Estimation. Procedia Computer Science 32, 356–363 (2014)
7. Chan, K.Y., Dillon, T.S., Singh, J., Chang, E.: Neural-Network-Based Models for Short-Term Traffic Flow Forecasting Using a Hybrid Exponential Smoothing and Levenberg–Marquardt Algorithm. IEEE Transactions on Intelligent Transportation Systems 13(2), 1524–9050 (2012)
8. Guo-jiang, S.: An intelligent hybrid forecasting model for short-term traffic flow. In: Proceeding of the 8th World Congress on Intelligent Control and Automation (WCICA), pp. 486–491 (2010)
9. Kamarianakis, Y., Prastacos, P.: Space–time modeling of traffic flow. Computers & Geosciences 31(2), 119–133 (2005)
10. Yuhrong, X., Liangzhog, J.: Water Quality Prediction Using LS-SVM and Particle Swarm Optimization. In: Second International Workshop on Knowledge Discovery and Data Mining, WKDD 2009, pp. 900–904 (2009)

11. Huang, Y.: Short-term traffic forecasting based on Wavelet Network Model combined with PSO. In: International Conference on Intelligent Computation Technology and Automation (ICICTA), vol. 1, pp. 249–253 (2008)
12. Arthur, S.: Some Studies in Machine Learning Using the Game of Checkers. IBM Journal 3(3), 210–229 (1959)
13. Vapnik, V.: The nature of statistical learning theory (2), pp. 8–23. Springer, New York (1998)
14. Luo, Z., Wang, P., Li, Y., Zhang, W., Tang, W., Xiang, M.: Quantum-inspired evolutionary tuning of SVM parameters. Progress in Natural Science 18(4), 475–480 (2008)
15. Carozza, M., Rampone, S.: Towards an incremental SVM for regression. In: Proceeding of the International Joint Conference on Neural Networks, Italy, pp. 405–410 (2000)
16. Mustaffa, Z., Yusof, Y., Kamaruddin, S.S.: Gasoline Price Forecasting: An Application of LSSVM with Improved ABC. Procedia - Social and Behavioral Sciences 129, 601–609 (2014)
17. Suykens, J.A.K., Van Gestel, T., De Brabanter, J., De Oor, B., Vanderwalle, J.: Least squares support vector machines. World Scientific Publishing Co. Pte. Ltd., Leuven (2002)
18. Yang, X.S.: Nature-inspired metaheuristic algorithms, 2nd edn. Luniver Press, United Kingdom (2010)
19. Yusof, Y., Kamaruddin, S.S., Husni, H., Ku-Mahamud, K., Mustaffa, Z.: Forecasting model based on LSSVM and ABC for natural resource commodity. International Journal of computer theory and Engineering 5(6) (2013)
20. Pelkmans, K., Suykens, J.A.K., Gestel, T.V., Barbanter, J.D., Lukas, L., Hamer, B., et al.: LSSVM: A Matlab/c toolbox for least squares support vector machines (2002), http://www.esat.kuleuven.be/sista/lssvmlab

Implementation of Dynamic Traffic Routing for Traffic Congestion: A Review

Norulhidayah Isa[1], Azlinah Mohamed[1], and Marina Yusoff[1,2]

[1] Faculty of Computer and Mathematical Sciences,
[2] Institute of Infrastructure Engineering and Sustainaibility Management,
University Technologi MARA
40450 Shah Alam, Selangor, Malaysia
`norul955@tganu.uitm.edu.my`, `{azlinah,marinay}@tmsk.uitm.edu.my`

Abstract. Traffic congestion is a condition where traffic demands exceed traffic capacity. It is a global problem in transportation that occurs around the world especially in metropolitan city. Dynamic traffic routing has been recognized as one of the methods that is capable of dispersing traffic congestions efficiently. This paper reviews the recent implementations of dynamic traffic routing in traffic congestion problems. Study on how the dynamic or online concept has been implemented in traffic routing focusing on definition of dynamic routing, traffic routing environment, traffic routing policy and routing strategy is reviewed in this paper. Some issues such as proactive routing and handling non-recurrent congestion are properly expounded while highlighting some limitations as well as suggestions for future research. As a conclusion, dynamic traffic routing is shown to be an important method in optimizing traffic congestion release. More studies need to be conducted in search of better solution.

Keywords: Traffic routing, Dynamic traffic routing, Online routing, Traffic Congestion.

1   Introduction

Traffic Congestion occurs in various network domains such as, airline travel-planning, video streams in computer networks and transportation network systems [33]. The problem happens when demand is higher than network capacity which leads to the disruption in traffic flow of the network. Traffic congestion in transportation domain has been recognized as a global issue especially in urban areas. One contributing factor to traffic congestion is the increase in human population. Traffic congestion is not only troublesome to drivers, but also lead to increase in air pollution [27] as well as higher probability of accidents [22],[26].

Traffic congestion can be divided into two categories namely recurrent and non-recurrent. Recurrent congestion is a cyclical basis or regular type of congestion due to imbalanced traffic flow during peak hours [16], complex network design structure and planning [47], and frequent ramp on and ramp off [36]. Non-recurrent congestion is an irregular network disruption caused by unplanned incidents and natural disasters

© Springer Science+Business Media Singapore 2015
M.W. Berry et al. (Eds.): SCDS 2015, CCIS 545, pp. 174–186, 2015.
DOI: 10.1007/978-981-287-936-3_17

[30], road constructions and work zones[41],[43], and specially planned events such as football tournaments [23].

Various solutions have been developed and implemented in solving traffic congestions especially for congestions in urban areas. Chen et al.[4] categorized the solutions into two strategies; hard and soft . Hard strategy involves changes in network topology structure while soft strategy focuses on managing the traffic based on current structure. Hard strategy will incur extra cost and manpower compared to soft strategy. In addition, Liang and Wakahara[24] categorized the solutions into three levels; solution that focuses on reducing traffic demands for example congestion pricing [13], shifting road traffic to other travel modes for example ride and share [1], and dispersing the traffic to maximize the usage of traffic network capacity such as Route Guidance System (RGS) [7],[20].The last solution is the most significant solution. This is due to the ability of reducing the demand especially during peak hours. This is because the first solution will not lessen non-recurrent congestion [15] and for the second solution, it is impossible to control people's mode of transportation.

Traffic routing is known for its capability to normalize the traffic flow by distributing the traffic demands efficiently throughout the network. There are two steps in traffic routing which are traffic route pre-planning and traffic re-routing. In route pre-planning, the idea is to provide route suggestions to the drivers before they left to the destination. Drivers are expected to follow the suggested routes throughout their journey. While in traffic re-routing, the suggested routes will be altered considering several factors, such as real time congestion information and possibilities of congestion. Thus, the vehicle will be re-routed to the least congested route.

This paper reviews the implementation of traffic routing especially dynamic traffic routing in dispersing the traffic in both recurrent and non-recurrent types of congestions. It will discuss how dynamic traffic routing has been implemented in current researches and their limitations.

2   Traffic Routing

Traffic routing is the process of improving traffic flow by redirecting and re-routing the vehicles in the traffic network according to changes in traffic conditions. The main purpose of traffic routing is to homogenize and improve traffic flow in network structure. It is a different problem compared with Vehicle Routing Problem (VRP). In VRP, the algorithm tends to find the shortest path for vehicles to visit all nodes at a time. While in routing process, the objective is to find the best route for vehicles from its origin to the destination.

The routing process is performed with several objectives. In earlier years of routing development, most of the researchers focused on finding the shortest travelling distance. However, with the development of traffic network and the increase in traffic routing process, it is acceptable to sacrifice distance over the time as long as it would prevent traffic breakdown in the network [40]. Regardless of time and distance consideration, the most important objective in solving congestion problems is to normalize the traffic flow. Traffic flow is defined as the number of vehicles passing a given location per time unit [40]. To disperse the traffic flow efficiently, traffic capacity must be fully utilized. Traffic capacity in transportation engineering is defined as the maximum hourly rate vehicles could travel from one point to another in a given time period [37]. That means traffic capacity can be represented by the number of vehicles per hour on a certain lane. To utilize the traffic capacity, traffic assignment must be manipulated where vehicles will be assigned to route depending on traffic state and time. This can be calculated through traffic density. Traffic density is expressed as the number of vehicles on a road segment at a given time [40]. It represents traffic flow over vehicle speed. Traffic flow is number of vehicles passing the specific point in a time interval. Thus, traffic density can be expressed as shown in equation 1.

$$\text{Traffic density} \quad = Q/V \tag{1}$$

Where: $Q = \frac{\Delta N}{\Delta T}$ (N: number of vehicle, T: time interval), V = Vehicle speed

According to Wardrop principle, improving traffic flow can be done according to two principles which are System Optimal (SO) and User Equilibrium (UE) [37]. SO principle will find the best possible network performance route. It will improve total travel time for whole network and sacrifice individual performance. While UE principle attempts to improve individual performance disregarding the total network performance.

3   Dynamic Traffic Routing

Traffic congestion is a dynamic problem where the traffic environments change continuously over time. Traffic congestion recovery, especially for non-recurrent congestion, can be done by detouring the traffic towards less congested area. In recent years, solutions like Advanced Traveler Information System (ATIS) and Intelligent Transportation System (ITS) have been widely used. The main component in both systems is Dynamic Route Guidance (DRG) which reallocates a new route in order to disperse users from congested lane towards less congested area. The core role of DRG is dynamic traffic routing process where the system must be able to capture real time information and compute the optimal path within acceptable computational time [20].

Dynamic routing falls under dynamic optimization problem where input elements of the problem are changing over time. The objective function for dynamic optimization problem can be deterministic at a given time; however it can be changed throughout optimization process [38]. Russel and Norvig[33] stressed that dynamic algorithm must process the input data as they are received rather than waiting for the input data set become available. In defining dynamic routing Treiber et al. [40] suggested that the routing process must consider the changes in traffic demand and network infrastructure. While Sever et al.[34] stressed that dynamic routing must update the route upon realization of disruption while travelling through the network. Thus, it can be concluded that there are two processes in dynamic routing. First is capturing the changes in the network environment which includes cost and constraints. Second is the process of updating the route or the algorithm responses towards the internal changes and feedbacks along the routing processes. These will be further discussed in Section 3.1 and 3.2.

3.1 Deterministic and Stochastic Environment

According to literature review, traffic routing was developed in two types of environments; deterministic and stochastic. Russel and Norvig [33] highlighted that the specification of these environments is crucial in designing the algorithm. In deterministic environment, current state will determine the next state of environment and action executed by algorithm [33]. The environment is fully observable where it uses static representation of network structure; variables and cost of the surface are predetermined and constant[10],[35]. However, most of the recent DRGs have been implemented in stochastic environments where travelling cost and changes in traffic environments are treated according to the dynamic nature of the problem. In stochastic environment, the next state is partially observable and non-deterministic. The changes will be captured and stored for routing purposes. Stochastic variables are varied from fluctuating cost between two nodes and the changes in the network topology. Most of the researcher considered the road cost as stochastic variables. However, only selected variables will be treated as stochastic.

To illustrate the implementation of stochastic environment, columns three and four in Table 2 depicts several stochastic variables and variables' update strategy in routing process. First, it shows that the most common stochastic variables that had been employed is route density [6],[7], [9],[19],[22],[44]. Route density represents the relation between traffic demand and route capacity. Second is traffic demand or amount [8],[24],[31] where number of vehicles on the road in a time is counted and updated. Other variables that have been employed are vehicle speed, and traffic flow. Traffic flow represents number of vehicle passing certain point on network in a time interval (usually in one hour [24]). Conventionally, route density captures the most researchers' attention in traffic condition representation, however there is an argument saying that traffic amount is better than route density since it was calculated based on data captured in a time interval, which not presenting the current information even it was updated in discrete time[24].

There are two common methods to update the stochastic variables. First is time based update method which can be divided into two fold; discrete time and interval time. In discrete time, the update time is specific. For example if the time t=5 seconds, the variable will be updated for every five seconds. Meanwhile, for interval time, data is updated according to time range for example time t is between 5 to 10, so any change in variable value will be captured and updated within that time. Second method is node or intersection based. The update process happens whenever vehicles arrive at new node or intersection. It will be based on vehicles' positions.

3.2   Online and Offline Routing Policy

Referring to the dynamic routing definition defined in previous section, routing algorithm should be able to react according to the internal changes and feedbacks along the routing processes. It means that even when the vehicles have begun to traverse the route, the algorithm must consider re-routing the path if there are changes in traffic cost. In implementing traffic routing, there are two types of routing policy which are online and offline routing [34]   and some of the researchers use terms such as static and dynamic routing [37].

In offline policy, the route is pre-planned before vehicles begin travelling according to specified origin and destination. The routing process stops when the vehicles leave the destination. Offline policy can be implemented in stochastic environment where variables or traffic cost is updated before the route is generated. Another category of offline policy is known as robust strategy that manipulates predicted data in routing process. The route is generated with the guideline of predicted data on the assumption that the projected data is 100% correct.

For Online routing policy, the algorithm tries its best to respond towards the changes and feed back in real time manner even when the vehicles have started traversing the route. To implement this, the route is preplanned beforehand; when the vehicles started to traverse the route, the algorithm will monitor the changes of the variables and will check for the need to re-route. Therefore, online routing must be executed in stochastic environment.

Table 1 demonstrates twelve publications from 2012 to 2014 that used dynamic or real time in the tittle. It shows that even though it uses dynamic or real time in the title, the routing policy used is not necessarily dynamic or online. The word dynamic in the tittle is referring to the stochastic environment used in the routing process. Going through the twelve publications, it can be construed that the implementation of dynamic or real time can be divided into two, dynamic in both environment and algorithm policy and dynamic in terms of environment only. Even with the usage of stochastic variables, the second type of implementation will lead to oscillation of traffic congestion at certain path because it disregards the dynamic state of traffic congestion. Even if it uses robust policy where the congestion is predicted, cost will increase because the projected data are only assumed 100% correct [37].

To give a clear view on how dynamic traffic routing have been implemented, columns five to seven in Table 2 presents a review on Routing Policy, Optimization Method, and Online Policy Implementation in recent traffic routing implementations. According to the literatures, the implementation of Online routing policy was done in two fold; re-route in a time interval and re-route at a node or intersection. For time interval implementation, routing algorithm will re-calculate the remaining route in specific time frame. If there is a better route, the vehicles will be re-routed. If not vehicles will follow the pre-calculated route. Second method is re-calculating the best route at each node or intersection. When the vehicles arrived at an intersection, algorithm will calculate the next movement for the vehicles. If there is a better route compared to pre-calculated route, the vehicles will be suggested to follow the new route.

Re-routing process happens in several conditions, (1)when routing algorithm predicts there would be a congestion for the next traversal lane [24], (2)based on level of congestion impact towards current route [32], (3)when there is delay in travel time [5] , and (4) when the density of the route exceeds route capacity. The re-routing process was done by focusing on the origin of journey and destinations of each vehicle that is; routes were calculated for single vehicles according to their specified destinations. Then during the process, rerouting will be implemented either for all vehicles or just focusing on selected vehicles only. The selection of the vehicle process will be based on the urgency which will be measured according to the impact of congestion or disruption towards the vehicles.

Table 1. Recent publication on Dynamic Routing.

Ref.	Publication Tittle	Routing Policy
[5]	Research on Dynamic Route Guidance for An Emergency Vehicle Considering the Intersection Delay	Online
[8]	Multiple Constrained Dynamic Path Optimization based on Improved Ant Colony Algorithm	Online
[24]	Real Time Urban Traffic Amount Prediction for Dynamic Route Guidance Systems	Online
[25]	Dynamic Route Guidance Algorithm Based on Improved Hopfield Neural Network and Genetic Algorithm	Offline
[22]	Dynamic Travel Path Optimization System Using Ant Colony Optimization	Online
[29]	Distributed Regret Matching Algorithm for a Dynamic Route Guidance	Online
[46]	Dynamic Route Choice Based on Prospect Theory & Online	Online
[2]	Dynamic Route Choice Based on Prospect Theory	Offline
	Real Time Vehicle Routes Optimization by Cloud Computing in The Principle of TCP/IP	
[34]	Dynamic Shortest Path Problems: Hybrid Routing Policies Considering Network Disruptions	Online
[39]	Real-time Vehicle Route Guidance Based on Connected Vehicles	Offline
[45]	Dynamic Route Guidance Using Improved Genetic Algorithms	Offline
[11]	Dynamic Routing Under Recurrent and Non-recurrent Congestion Using Real-Time ITS Information	Online

3.3   Reactive versus Proactive

Another aspect that has been discussed in dynamic routing is type of data. In traffic routing, there are three categories of data commonly used. First category is historical data. Second category is data collected in real time representing current conditions of traffic environment and third category is predicted data, calculated based on current and historical data. Routing process that employs current data, or current data together with historical data is known as reactive routing. While routing process that include predictive data is known as predictive or proactive routing.

Reactive routing uses the snapshot of current traffic information to develop the route while proactive or also known as predicted routing strategy utilizes predicted traffic information to predict the future traffic condition. The implementation of the route planning in reactive strategy is faster [18], nevertheless it will lead to oscillation congestion problem since the future condition of the traffic is not considered [44].

In proactive routing, the routing is calculated considering future conditions. Different proactive variables have been used for example route density, time arrival probability, and route disruption possibility. The proactive data is used especially in avoiding congestion. References [6], [9], [44] consider route density of the routes based on on-going plan route in routing the vehicles. The constructed route will avoid the routes that have higher predicted route density. While Chen et al. [3] estimated probability of arriving at the destination based on expected travel time.

Table 2. Recent Traffic Routing Implementation

Ref.	Stochastic Variables	Update Strategy	Routing Policy	Optimization Method	Online Policy Implementation
[5]	Vehicle Speed	Intersection	Online	Dijkstra Algorithm	Time Interval
[7]	Route Density	Discrete Time	Offline	Inverse Ant Based System	NA
[8]	Traffic Amount	Discrete Time	Online	Genetic Algorithm	Intersection
[9]	Route Density	New Demand	Offline	Genetic Algorithm	NA
[14]	Traffic flow	Discrete Time	Online	Mathematical Programming	Intersection
[17]	Travel speed	Intersection	Online	Ant Colony Optimization	Time Interval
[22]	Route density	Intersection	Online	Ant Colony Optimization Algorithm	Intersection
[24]	Traffic Amount	Interval Time	Online	Dijkstra Algorithm	Travel Delay Exist or predicted
[29]	Traffic flow	Interval Time	Online	Regret Matching Algorithm	Intersection
[31]	Traffic Demand	Interval Time	Online	Ant Colony Optimization & Dijkstra Algorithm	Time Interval
[42]	Route Density	Discrete Time	Offline	Ant Colony Optimization	NA
[21]	Route Density	Intersection	Online	Brownian Agent	Intersection
[32]	Vehicle speed	Interval Time	Online	Dijkstra & A* Algorithm	Time Interval
[34]	Disrupted Lane	Discrete Time	Online	Backward Recursive Algorithm	Intersection
[39]	Vehicle Speed	Segmentation	Offline	Dijkstra Algorithm	NA
[44]	Route Density	New Available data	Offline	A* Algorithm	NA
[11]	Vehicle velocity, incidents & delays	Discrete Time	Online	Markov Decision Process	NA

3.4 Non-recurrent Congestion

Non-recurrent Congestion (NRC) is a traffic congestion due to unusual event and other factors that change the normal traffic condition. It will cause greater congestion compared to daily problem. A change of 5% in daily condition can be classified as

NRC [28], [12]. The changes are obtained from percentage lane occupancy, traffic volume and vehicle speed [12]. Non-recurrent congestion contributed up to 60% delay in United State [28] and caused 50% of traffic delay [11] in transportation domain. However, this problem attracts less attention compared to recurrent congestion due its infrequent occurrence [12].

Table 3 below depicts several literatures that consider non-recurrent congestion in their routing process. Literatures except for reference [28] re-route vehicles with the objective of avoiding the affected place. The solutions are focusing on routing the vehicles with the objective to avoid the congestion area. The congestion will be predicted according to delay. Even those researchers consider non-recurrent congestion in rerouting process, most of them provide planning to avoid the congestion but not consider vehicles that already trapped in the traffic. In addition, route availability is neglected. Route availability is one of the most important in re-routing process [4] especially when it involved total route closure.

Table 3. Non-recurrent Congestion(NRC) in Dynamic Routing

Author	Variables to Consider	NRC	NRC Handling Methods
[17]	Travel delay	Difference between current & regular delay	Re-route each vehicles(avoid low speed route).
[28]	Flow rate, incident duration & blocked lane info.	5% changes from normal condition	Provide detour plan for traffic control management
[8]	Route Vulnerability.	Predicted based on current and historical data	Vulnerable lane will be avoided for each vehicle
[11]	Incident delay	Incident delay is predicted using Markov Chain Model	The vehicles will be re-routed based on new travel time

4   Findings and Discussions

In summary, there are three main components in traffic routing implementation; i) routing environment, ii) routing policy, and iii) update strategy for environment and routing policy. Dynamic traffic routing falls under online routing policy and must be implemented in stochastic routing environment. To adapt with the changes, the variables and routes must be updated frequently. Update strategy will specify the method uses in updating routes and variables. These finding is shown in Figure 1 above. In real implementation, the route will be updated in specified time; however the implementation of new route will be employed at the next intersections or nodes due to nature of network structure.

In developing dynamic traffic routing, stochastic environment is used to interpret the real environment situation into computer readable. Basic representation for city network structure can be represented using Graph set where $G(A, N)$ where N is a set of nodes and A is a set of links. Then traffic data must be incorporated with the city network structure. Both can be done through several method such as 3D simulation software for example StarLogo Software [5], Traffic Model Simulation software MAINS2IM (Multi- modAl INnercity SIMulation) [7] and others. Simulated data [5]

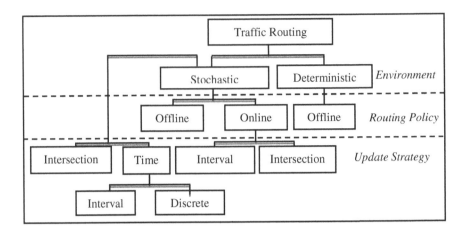

Fig. 1. Traffic Routing Conceptual Framework

or real scenario will be integrated into the simulation software to represent real network structure. The real data were collected uses several methods such as through Vehicle Adhoc Network, Geographical Information System, and loop detector. The data is collected and must be updated to the simulation environment. The update strategy for the data consists of two types. First, the data is updated every time vehicles arrived at the intersection. Vehicles acted as an agent to collect data such as vehicle's location and travel time along. Every time vehicles arrived at the intersection, collected data will be updated. Second strategy is time based strategy. It can be divided into two which are discrete and interval time strategy.

In dynamic traffic routing, online routing policy must be implemented. The objective of online routing policy is to adapt with the changes in traffic routing environment. This will involve constructing a new route for vehicles. Researchers used traffic density as an indicator in constructing new route. The process can be done two types of update strategy. First is in time interval and second is when the vehicles arrived at new intersection. Balanced route will be re-evaluated, if high density lane is detected in pre-planned route, the pre-planned route will be re-evaluated. New route will be calculated and cost for both routes (pre-planned and new route) will be compared. If new route has a better travel cost, vehicles will continue their journey with new route, if not it will use the pre-planned route.

Dynamic traffic routing has been extensively study and proven as one of the methods to reduce traffic congestion. Fluctuating cost between nodes in terms of route density and traffic flow have been extensively used in developing the dynamic traffic routing. These variables have been proven able to represent current vehicle movement on specific path and could be used in solving traffic congestion and travel delay. However it is not sufficient in capturing the non-recurrent congestion especially when there involves route closure or route availability. Uncertain event like flash flood or fatal accident will lead to serious congestion and will involve several road closures. This problem requires detail solution in detour the vehicles at the congested area

Among the related issues highlighted by researchers in traffic routing are the algorithm's capability in providing real time guidance and the congestion handling efficiency [32]. Most of the proposed routing algorithm assumed the origin and destination of each vehicle are known beforehand. For a large scale network, it would be computationally expensive to calculate or re-route all the vehicles according to the origin and destination during congestion [32]. Current solutions proposed by researchers are to select certain most affected vehicles to be re-routed[24], [32]. [24] found that vehicles which were not involved in re-routing would benefit more in terms of travelling distance compared to vehicles involved in re-routing. Besides, increase in number of re-routings would increase the computational cost. Research findings also show at least 15 percent of the vehicles involved in re-routing will have longer travelling time compared to vehicles which were not [32].

5   Conclusion

In this paper, recent publications in dynamic traffic routing for traffic congestion were reviewed. This review suggests that in general, Traffic Routing Conceptual Framework can be divided into three levels as shown in Figure 1. This framework constitute to the field of traffic routing where it can be used as a reference for further research.

Although the issue of enhancing network performance using dynamic traffic routing has been extensively study, there are a couple of limitations that need to be looked into. One of them is most of the developed researches were focusing on guiding the specific vehicle or single traveler route guidance based on specific origin and destination. Routes are calculated according to the each vehicle preferences. Even though some of the developed route guidance was based on system optimum traffic assignments they still concentrate on individual vehicles. Each vehicle will be guided towards less congested area according to each vehicle preferences.

In disseminating traffic congestion especially in non-recurrent congestion, the main objective is to reduce the impact of the congestion for the whole network by dispersing the traffics at the affected area and to avoid incoming traffic from flowing to the affected area. The setback is, to apply dynamic route guidance for the whole route will require a lot of computational time especially when it involves various origins and destinations. So the question to be answered is what if instead of focusing on single vehicle the dynamic route guidance focuses on multiple vehicles; the objective is to withdraw vehicles from the congested area. So, instead of re-route the vehicles according to their specific preferences, detour plan should be provided to bring the vehicles to the nearest less congested area. The plan should consider all possible directions nearby in constructing the routes.

In non-recurrent congestion, a pre-planned road closure must be executed due to various reasons for example constructions, Independence Day ceremony and etc. In this situation, route density cannot be used in the routing process. In Kuala Lumpur, Malaysia, current implementation is by informing road users beforehand but with no proper guideline given. This raises another question to be answered, what is the best method to re-route the vehicles in such situation.

References

1. Agatz, N., et al.: Optimization for dynamic ride-sharing: A review. European Journal of Operational Research 223(2), 295–303 (2012)
2. Cai, M., et al.: Realtime vehicle routes optimization by cloud computing in the principle of TCP/IP. In: 2013 10th International Conference on Service Systems and Service Management (ICSSSM), pp. 113–118 (2013)
3. Chen, B.Y., et al.: Finding reliable shortest paths in road networks under uncertainty. Networks and Spatial Economics 13(2), 123–148 (2013)
4. Chen, S., et al.: Traffic dynamics on complex networks: a survey. Mathematical Problems in Engineering 2012 (2011)
5. Chen, S.S., et al.: Research on Dynamic Route Guidance for an Emergency Vehicle Considering the Intersection Delay. Applied Mechanics and Materials, 848–852 (2014)
6. Cong, Z., et al.: Ant Colony Routing algorithm for freeway networks. Transportation Research Part C: Emerging Technologies 37, 1–19 (2013)
7. Dallmeyer, J., et al.: Don't go with the ant flow: Ant-inspired traffic routing in urban environments. Journal of Intelligent Transportation Systems (2014) (just-accepted)
8. Dewen, S., et al.: Multiple Constrained Dynamic Path Optimization based on Improved Ant Colony Algorithm. International Journal of U- & E-Service, Science & Technology 7(6) (2014)
9. Dezani, H., et al.: Optimizing urban traffic flow using Genetic Algorithm with Petri net analysis as fitness function. Neurocomputing 124, 162–167 (2014)
10. Gubins, S., Verhoef, E.T.: Dynamic bottleneck congestion and residential land use in the monocentric city. Journal of Urban Economics 80, 51–61 (2014)
11. Güner, A.R., et al.: Dynamic routing under recurrent and non-recurrent congestion using real-time ITS information. Computers & Operations Research 39(2), 358–373 (2012)
12. Gurupackiam, S., et al.: A Snapshot of Lane-specific Traffic Operations under Recurrent and Non-recurrent Congestion. International Journal of Traffic and Transportation Engineering 3(4), 199–205 (2014)
13. Hoffman, K., et al.: Congestion pricing applications to manage high temporal demand for public services and their relevance to air space management. Transport Policy 28, 28–41 (2013)
14. Hu, P., Ma, Z.: Travel Efficiency in Urban Traffic Networks Based on Routing Strategies. Journal of Applied Science and Engineering Innovation 1(5) (2014)
15. Isa, N., et al.: A review on Recent Traffic Congestin Relief Approaches. In: 4th International Conference on Artificial Intelligence with Applications in Engineering and Technology. IEEE (2014)
16. Ishikawa, K., et al.: A decision support model for traffic congestion in protected areas: A case study of Shiretoko National Park. Tourism Management Perspectives 8, 18–27 (2013)
17. Jabbarpour, M.R., et al.: Ant-based vehicle congestion avoidance system using vehicular networks. Engineering Applications of Artificial Intelligence 36, 303–319 (2014)
18. Jahn, O., et al.: System-optimal routing of traffic flows with user constraints in networks with congestion. Operations Research 53(4), 600–616 (2005)
19. Jiang, B., Xu, X., Yang, C., Li, R., Terano, T.: Solving Road-Network Congestion Problems by a Multi-objective Optimization Algorithm with Brownian Agent Model. In: Corchado, J.M., et al. (eds.) PAAMS 2013. CCIS, vol. 365, pp. 36–48. Springer, Heidelberg (2013)

20. Jiang, B., et al.: Time-dependent pheromones and electric-field model: a new ACO algorithm for dynamic traffic routing. International Journal of Modelling, Identification and Control 12(1), 29–35 (2011)
21. Jiang, Z., Li, S.: Research on Optimized Control Model of Freeway Based on Dynamic Traffic Demand Estimation. Advances in Mechanical Engineering 2014 (2014)
22. Kponyo, J., et al.: Dynamic Travel Path Optimization System Using Ant Colony Optimization. In: Proceedings of the 2014 UKSim-AMSS 16th International Conference on Computer Modelling and Simulation, pp. 142–147 (2014)
23. Kwoczek, S., et al.: Predicting Traffic Congestion in Presence of Planned Special Events (2014)
24. Liang, Z., Wakahara, Y.: Real-time urban traffic amount prediction models for dynamic route guidance systems. EURASIP Journal on Wireless Communications and Networking 2014(1), 1–13 (2014)
25. Lin, N., Liu, H.: Dynamic route guidance algorithm based on improved hopfield neural network and genetic algorithm. Int. J. Innov. Comput., Inf. Control. 10(2), 811–822 (2014)
26. Litman, T.: Factors to Consider When Estimating Congestion Costs and Evaluating Potential Congestion Reduction Strategies (2013)
27. Litman, T.: Smarter Congestion Relief in ASIAN Cities. Transport and Communications Bulletin for Asia and the Pacific 82(1) (2013)
28. Liu, Y., et al.: Decision Model for Justifying the Benefits of Detour Operation under Non-Recurrent Congestion. Journal of Transportation Engineering 139(1), 40–49 (2013)
29. Ma, T.-Y.: Distributed Regret Matching Algorithm for a Dynamic Route Guidance. In: Jezic, G., Kusek, M., Lovrek, I., Howlett, R.J., Jain, L.C. (eds.) Agent and Multi-Agent Systems: Technologies and Applications. AISC, vol. 296, pp. 107–116. Springer, Heidelberg (2014)
30. Mun, L.S.: Blue Ocean Strategy in Traffic Management for Bandaraya Kuala Lumpur (2013)
31. Ochiai, J., Kanoh, H.: Hybrid Ant Colony Optimization for Real-WorldDelivery Problems Based On Real Time and Predicted Traffic in Wide Area ROad Network. Computer Science (2014)
32. Pan, Y., Xu, J.: Traffic network design problem with uncertain demand: A multi-stage bi-level programming approach. World Journal of Modelling and Simulation 9(1), 68–73 (2013)
33. Russell, S., Norvig, P.: Artificial intelligence: A modern approach. Prentice Hall Pa. (2009)
34. Sever, D., et al.: Dynamic shortest path problems: Hybrid routing policies considering network disruptions. Computers & Operations Research 40(12), 2852–2863 (2013)
35. Spears, W.M., Prager, S.D.: Evolutionary search for understanding movement dynamics on mixed networks. GeoInformatica 17(2), 353–385 (2013)
36. Spiliopoulou, A., et al.: Macroscopic traffic flow model validation at congested freeway off-ramp areas. Transportation Research Part C: Emerging Technologies 41, 18–29 (2014)
37. Suson, A.C.: Dynamic routing using ant-based control (2010)
38. Talbi, E.-G.: Metaheuristics: from design to implementation. John Wiley & Sons (2009)
39. Tian, D., et al.: Real-Time Vehicle Route Guidance Based on Connected Vehicles. In: Green Computing and Communications (GreenCom), 2013 IEEE and Internet of Things (iThings/CPSCom), IEEE International Conference on and IEEE Cyber, Physical and Social Computing, pp. 1512–1517 (2013)
40. Treiber, M., Kesting, A.: Traffic Flow Dynamics: Data, Models and Simulation. Springer, Heidelberg (2013) ISBN 978-3-642-32459-8

41. Tympakianaki, A., et al.: Real-time merging traffic control for throughput maximization at motorway work zones. Transportation Research Part C: Emerging Technologies 44, 242–252 (2014)

42. Wei, C., et al.: Formulating the within-day dynamic stochastic traffic assignment problem from a Bayesian perspective. Transportation Research Part B: Methodological 59, 45–57 (2014)

43. Weng, J., Meng, Q.: Estimating capacity and traffic delay in work zones: An overview. Transportation Research Part C: Emerging Technologies 35, 34–45 (2013)

44. Wilkie, D., et al.: Adaptive Route Planning for Metropolitan-Scale Traffic. In: SPARK 2013 (2013)

45. Yu, Z., et al.: Dynamic route guidance using improved genetic algorithms. Mathematical Problems in Engineering 2013 (2013)

46. Zhang, W., He, R.: Dynamic Route Choice Based on Prospect Theory. Procedia-Social and Behavioral Sciences 138, 159–167 (2014)

47. Zhao, D., et al.: Computational intelligence in urban traffic signal control: A survey. IEEE Transactions on Systems, Man, and Cybernetics, Part C: Applications and Reviews 42(4), 485–494 (2012)

Part IV
Pattern Recognition

A Comparative Study of Video Coding Standard Performance via Local Area Network

Siti Eshah Che Osman[1], Hamidah Jantan[2], Mohamad Taib Miskon[3],
and Wan Ahmad Khusairi Wan Chek[3]

[1] Faculty of Computer and Mathematical Sciences, Universiti Teknologi MARA,
40450 Shah Alam, Selangor, Malaysia
cteshahco@gmail.com
[2] Faculty of Computer and Mathematical Sciences, Universiti Teknologi MARA,
23000 Dungun, Terengganu, Malaysia
hamidahjtn@tganu.uitm.edu.my
[3] Faculty of Electrical Engineering, Universiti Teknologi MARA,
23000 Dungun, Terengganu, Malaysia
{moham424,wanah5079}@tganu.uitm.edu.my

Abstract. In order to ensure the compatibility among video codecs from distinct manufacturers and applications, intensive efforts have been undertaken in recent years. For example, a digital video, its size is very large to be stored in memory of storage device. Practically, video should be processed to make it more practical to be shared, at the same time maintaining the quality of the video and avoiding error rate to occur during the transmission. All the issues were discussed in this paper. This paper describes the comparison of video coding standard and discusses on video transmission. For instance, a sample video has 320 x 240 pixels per frame, 24 frames per second, total 265 minutes full color video. Thus, several video compression standards had been used to analyse the throughput and round trip time performance base on different bit rates. The result shows the video with a higher bit rate will have a higher throughput. This experiment could proceed by applying a new method of video compression with the latest video coding standard to analyse the performances.

Keywords: Video Coding Standard, Video Transmission, Bit Rate, Throughput, Round Trip Time.

1 Introduction

In the late 1980's and 1990's, video compression became the main the area of research and it enabled a diversity of applications which included video broadcast, video conferencing and many more [1]. Transferring a big size of data is a common problem to happen [2]. Normally, a big file will occupy a space and it will become harder to be transmitted. Compression is needed to solve this problem by reducing the file size at the same time maintaining the quality of an original file.

© Springer Science+Business Media Singapore 2015
M.W. Berry et al. (Eds.): SCDS 2015, CCIS 545, pp. 189–197, 2015.
DOI: 10.1007/978-981-287-936-3_18

The main goal of most digital video coding standards has been to optimize coding efficiency and there are various video coding techniques proposed and many researches still ongoing out there [3]. Coding efficiency is the ability to reduce the bit rate necessary for representation of video content to reach a given level of video quality. Its aim is to maximize the video quality achievable within a given available bit rate compared to the existing standards [4].

Video streaming based on Transport Control Protocol (TCP) has become popular because it is easy in handling and deploying [5]. TCP provides error recovery by requesting retransmission of missing data and without error recovery, every time a packet is lost during transmission, there would be an audio or video glitch in the playback. Besides, in low latency networks, TCP features good throughput performance and low end-to-end delays, that make TCP-based interactive services possible.

Through the previous study, a new video compression standard for a very low bit rate has been discovered. This paper is aimed to examine the performance of the throughput and round trip time when different coding standards of a file were used to transmit over wired-LAN by two PCs. This preliminary study demonstrates that the throughput is higher at a low bit rate, both for sending and receiving a file. This experiment will help any continuous project in the future to obtain a better performance in the transmission process.

This paper is discussed as follows; the second section explains the related work on the video coding standard. The third section discusses the video process and transmission continues with the video performance testing for the next section. Then, the next section explains about the experiment setup. The sixth section discusses the result and discussion conducted in this study. Finally, this paper ends with the conclusion remarks and future enhancements.

2   Video Coding Standard

Video coding standards have been developed primarily through the development of the well-known ITU-T (International Telecommunication Union-Telecommunication) and ISO/IEC (International Organization for Standardization/International Electrotechnical Commission) standards [6]. The Video Coding Experts Group (VCEG) is a working group of the ITU-T, which was formed in 1984 and was responsible for the H.26x video coding standards. The Moving Picture Experts Group (MPEG) was formed in 1988, and is a working group of experts from ISO/IEC targeted at developing standards for digital audio and video coding and transmission [7, 8].

The H.263 [7] standard was developed by the ITU-T VCEG in 1995. It was based on H.261, MPEG-1 and MPEG-2. The aim for this standard is to operate at a low bitrate for video conferencing. Besides, it can offer better quality at all bitrates in comparison with the prior standards. The original standard was improved in 1998 and 2000, and the coding efficiency and capacities were considerably enhanced. There have been many new features developed, for instance, reference picture selection mode, support for flexible picture formats and modified quantization mode.

In 1998, the MPEG-4 visual standard was developed by ISO/IEC MPEG and aimed at low bit rate video communications [7]. The standard was extended later from a low bit rate up to a high one for more efficient work. In comparison to MPEG-2, the coding efficiency was enhanced. In addition, some new features were added in the standard, such as error resilience and segmented coding of shapes. Besides, during the development in 2000 and 2001, more features were added for example, variable block size motion compensation, intra discrete cosine transform coefficient prediction and quarter-sample motion compensation.

The year 2003, when the first specifications of the H.264/AVC was approved [9], can be regarded as another milestone. H.264 is also called MPEG-4 Part 10 and AVC that was developed jointly by the ITU-T VCEG and ISO/IEC MPEG, is known as Joint Video Team (JVT) [7]. It has been an enabling technology for digital video in almost every area that was not previously covered by H.262/MPEG-2 Video and has considerably displaced the older standard within its existing application domains. Currently, this coding scheme is used in most of applications like Digital Video Broadcasting (High Definition) TV signals over satellite, Blu-ray Discs, cable and terrestrial transmission systems, camcorders,   security applications, video recording on mobile phones, video telephony, on-line video streaming [9, 10].   H.264 coding offers higher quality for lower bitrates than MPEG-2 and is also more suitable for HD broadcasting and screening. There were several new features included in the standard, for instance, the number of reference pictures was increased up to 16 frames, more flexible variable block-size motion compensation was used with the block sizes as small as 4×4 and as large as 16×16 besides quarter-pixel motion compensation was included. In addition, two main new features were added to the standard, which are Scalable Video Coding (SVC) and Multiview Video Coding (MVC). The SVC extensions were completed in 2007 and the MVC extensions were completed in 2009.

As the demands of video quality at lower bandwidth are increasing, a new video coding standard is needed. The most latest standard of the ISO/IEC Moving Picture Experts Group (MPEG) and ITU-T VCEG is High Efficiency Video Coding (HEVC), which is also known as MPEG-H Part 2 or ITU-T H.265 [9]. The HEVC standard was completed in January 2013. It can achieve a bit rate reduction of 50% for equal perceptual video quality, and can support increased video resolution such as ultra-high definition television [7].

As shown in the Fig. 1 below, the graph compares bit rate levels across MPEG-2, H.264/AVC & H.265/HEVC. The improvement has been made to provide enough flexibility to allow the standard to be applied to a wide variety of applications on a wide variety of networks and systems. The graph shows the target of bit rates for every new video standard will be decreased until 50%. Table 1 below shows the major comparison between the two common standards.

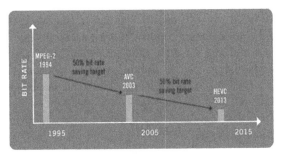

Fig. 1. Graph Comparing Bit Rate Levels Across MPEG-2, AVC & HEVC [11]

Table 1. Major Comparison between H.264/AVC and H.265/HEVC.

	H.264/AVC	H.265/HEVC
Max Bit rate (Mbps)	25	50% bit rate reduction
Picture Rate (fps)	59.94	Up to 300
Resolution standard	Up to 4k	Up to 8k

3   Video Process and Transmission

Codec is software with two components; an encoder and a decoder. The former takes a raw data files and changes it into a compressed file. This allows a more efficient storage and transmission of the data. The latter is the inverse process to restore the original file [12]. Software and/or hardware that can encode and decode are called codecs [13, 14]. Video content that is compressed by using one standard cannot be decompressed with a different standard. Video codecs that implement different standard are normally not compatible with each other, for example, an MPEG-4 Part 2 decoder will not work with an H.264 encoder [15]. This is simply because one algorithm cannot correctly decode the output from another algorithm but it is possible to implement many different algorithms in the same software or hardware, which would then enable multiple formats to be compressed.

There is a wide range of video communication and streaming applications with different operating conditions or properties. For example, a video communication application may be for point-to-point communication, multicast or broadcast communication [1]. Different network protocols offer different services and there are different video streaming methods. These methods can be counted as User Datagram Protocol (UDP), Transport Control Protocol (TCP), Real-Time Transport Protocol (RTP), Real-Time Streaming Protocol (RTSP), Hypertext Transfer Protocol (HTTP) [16].

For this experiment, we used streaming media over TCP. By using TCP, it may seem natural to employ because of the given success and ubiquity of TCP. Indeed, there are several important advantages of using TCP. Firstly, TCP rate control has been proven for its reliability, stability and scalability [17]. Besides, it guarantees delivery via retransmissions and acknowledgements [18]. It provides a mechanism for reducing the transmission rate in the presence of network congestion, which reduces overall network packet loss rates, and hence the need for retransmissions. However, there is a difficulty in using TCP for streaming video delivery which is media delivery guarantee of TCP is accomplished through persistent retransmissions with potentially increasing waiting time and also delivery time.

4   Video Performance Testing

In most case studies, the current video coding standards have been discussed generally in terms of the features and also their characteristics [19, 20]. Besides, the quality measurement, particularly peak signal-to-noise ratio (PSNR) over bit rate, has also been discussed [4, 21, 22]. All generations of video coding standards were designed to provide improved compression efficiency and increased their functionality [1]. Compression efficiency is the main attribute for a video codec to be successful in wireless environments [23]. In a network testing, there are several common aspects required for video visualization which include bandwidth, threshold, throughput, frame rates and also screen resolution [20]. Those are common aspects that have to be taken into consideration in order to obtain good performance as well as user satisfaction. Besides the measurement of the video quality, performance of transmitting 3D video over LTE network had been discussed in the previous paper [24] which aimed to deploy the 3D video services over LTE networks with Quality of Service (Qos) with limited numbers of users. An evaluation of the throughput over number of users was also measured. This paper discovers the efficiency of transmitting a video file with three different video coding standards over Wired LANs network that only involved two PCs in order to analyse the throughput and also the round trip time.

5   Experiment Setup

This experiment consisted of three phases. Fig. 2 shows a framework phases which include the process involved in doing this experiment. The first phase is to gather information in order to understand the video coding standard in general and set-up control parameters of a different video coding standard. The second phase is the experiment set up. Fig. 3 shows the transmission process involved in general. For this experiment, we used unicast which is point-to-point communication between the client and server that required two PCs by using Transport Control Protocol (TCP). Once the connection succeed, the transmission started with the client starts to request a file from server. Then, TCP ensured that the client was ready to accept data. The TCP connection was established successfully via a three-way handshake. It responded with an ACK and SYN bits set. TCP made sure that the data reached its destination. If the receiver did not acknowledge a particular packet, it retransmitted the packet automatically up to three times. The final phase is the result analysis. For this phase, Wireshark network analyser was used to capture the network traffic. This phase analysed the throughput and the round trip time performance by using Wireshark network analyser.

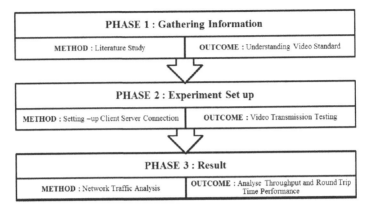

Fig. 2. The Framework Phases

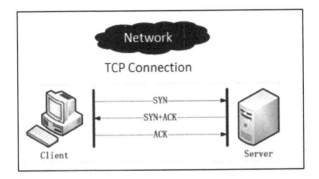

Fig. 3. Transmission Process

6   Result and Discussion

Table 2 below shows the parameters of three videos that were used to test the transmission performance by using Wired LANs connection. Only two PCs were used in this experiment. This phase was started with the setting up IP Address of the server in the client site. The result for this paper focuses on the performance test based on different video coding standards that were played consecutively. The first evaluation for this experiment was throughput performance, as shown in Fig. 4 and Table 3 below. Throughput was analysed based on the average rate of a successful message delivered over communication channel that involved a sender and receiver. Based on the graph performance, it shows that the video coding standard of H.264 has the highest throughput. A file with the highest bit rate tends to have the highest throughput. A few gaps are shown in the graph because of there is very little traffic going from the sender to the receiver during the transmission.

Table 2. Parameters of Three Videos.

Parameter \ Video Coding Standard	MPEG-2	H.264	H.265
Duration	4mn 25s	4mn 25s	4mn 25s
Overall Bitrate(Kbps)	396	356	228
File Size (MiB)	12.5	11.3	7.23
Screen Resolution	320x240	320x240	320x240
Frame Rate (fps)	24	24	24

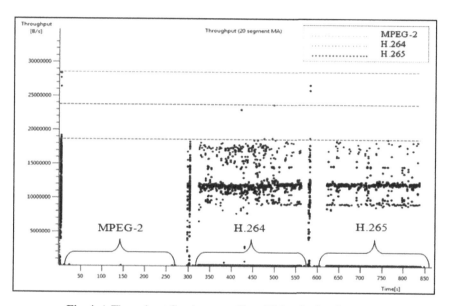

Fig. 4. A Throughput Graph among Three Video Coding Standards

Table 3. Result of Throughput.

Video Coding Standard	Throughput (B/s)
MPEG-2	28500000
H.264	23750000
H.265	18500000

Besides throughput, a round trip time performance has also been analysed in this experiment. The round trip time means the length of time it takes for a signal to be sent plus the length of time it takes for an acknowledgment of that signal to be received. Fig. 5 shows a round trip time graph among three video coding standards that was captured in the transmission process consecutively. Based on the result of a round trip time graph, H.265 shows the shortest sequence number as compared to the others.

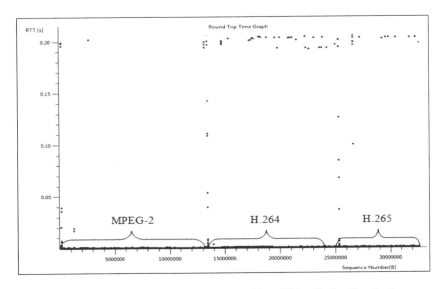

Fig. 5. A Round Trip Time Graph among Three Video Coding Standards

7   Conclusion and Future Work

As a conclusion, a bit rate is a property of video while throughput is a property of transmission. Currently, TCP is the most commonly used transport-layer protocol for a small transaction. It offers a reliable byte stream of retransmissions and congestion avoidance and it meets the requirement that so many applications desire. To optimize TCP throughput, the sender should send enough packets to fill the logical pipe between the sender and receiver. Generally, there is an improvement in video compression standard in accordance with the latest technological and technique developments. Based on the experiment, it shows that a file with a high bit rate would result in a high throughput. Besides, based on the result of round trip time graph, it shows that H.265 has small sequence number as compared to the other two video coding standards. The aim of this experiment is achieved, which focuses on the throughput and also round trip time. In future, this experiment could proceed by applying a new method of video compression with the latest video coding standard to analyse the performances.

Acknowledgements. This work has been supported by RAGS/2013/UITM/ICT05/1.

References

1. Apostolopoulos, J.G., Tan, W.T., Wee, S.J.: Video Streaming: Concepts, Algorithms, and Systems, HP Laboratories, report HPL-2002-260 (2002)
2. Sashikala, et al.: A Survey of Compression Techniques. International Journal of Recent Technology and Engineering (IJRTE) 2(1), 152–156 (2013)
3. Rao, K.R., Kim, D.N., Hwang, J.J.: Video coding standards: AVS China, H.264/MPEG-4 PART 10, HEVC, VP6, DIRAC and VC-1. Springer (2013)

4. Ohm, J.R., et al.: Comparison of the Coding Efficiency of Video Coding Standards—Including High Efficiency Video Coding (HEVC). IEEE Transactions on Circuits and Systems for Video Technology 22(12), 1669–1684 (2012)
5. Aggoun, A.: A 3D Dct Compression Algorithm For Omnidirectional Integral Images. In: IEEE International Conference on Acoustics, Speech and Signal Processing, Toulouse (2006)
6. Choi, S., et al.: Secure Video Transmission on Smart Phones for Mobile Intelligent Network. International Journal of Security and Its Applications 7(1), 143–154 (2013)
7. Zhang, G.: Computational Complexity Optimization on H.264 Scalable/Multiview Video Coding. In: School of Computing, Engineering and Physical Sciences, University of Central Lancashire, p. 124 (2014)
8. The Moving Pictures Group, http://mpeg.chiariglione.org/
9. Sullivan, G.J., et al.: Overview of the High Efficiency Video Coding (HEVC) Standard. IEEE Transactions on Circuits and Systems for Video Technology 22(12), 1649–1668 (2012)
10. Zach, O., Slanina, M.: A Comparison of H.265/HEVC Implementations. In: 56th International Symposium ELMAR (ELMAR), Zadar (2014)
11. Grotticelli, M.: Industry Reacts To HEVC License Fees.
    https://www.thebroadcastbridge.com/content/entry/899/
    industry-reacts-to-hevc-license-fees
12. Ozer, J.: Streaming 101: The Basics - Codecs, Bandwidth, Data Rate and Resolution,
    http://www.streaminglearningcenter.com/articles/streaming-
    101-the-basics–codecs-bandwidth-data-rate-and-
    resolution.html
13. Siegchrist, G.: Codec. http://desktopvideo.about.com/od/glossary/g/
    codec.htm
14. Bhojani, D.R., Dwivedi, D.V.V.: Novel Idea for Improving Video Codecs. International Journal of Electronics and Communication Engineering & Technology (IJECET) 4(2), 301–307 (2013)
15. Kaur, P., Sharma, E.S., Ahuja, E.S.P.S.: Latest Video Compression Standard H.264 Within Video Surveillance. International Journal of Advanced Research in Computer Science and Software Engineering 2(1) (2012)
16. Zerman, E.: A Portable Stereo-video Streaming System. In Electrical and Electronics Engineering Department, Middle East Technical University, p. 90 (2013)
17. Kima, T., Ammarb, M.H.: Receiver Buffer Requirement for Video Streaming over TCP. Visual Communications and Image Processing, vol. 6077 (2006)
18. Kuschnig, R., Kofler, I., Hellwagner, H.: Improving Internet Video Streaming Performance by Parallel TCP-based Request-Response Streams. In: 7th IEEE Consumer Communications and Networking Conference (CCNC), Las Vegas (2010)
19. Ponlatha, S., Sabeenian, R.S.: Comparison of Video Compression Standards. International Journal of Computer and Electrical Engineering 5(6), 549–554 (2013)
20. Junghare, U.S., Sherekar, V.M.T.D.S.S., Dharaskar, R.V.: Compression Techniques for Remote Visualization on Mobile Platforms. In: National Conference on Innovative Paradigms in Engineering & Technology, NCIPET (2012)
21. Grois, D., et al.: Performance Comparison of H.265/MPEG-HEVC, VP9, and H.264/MPEG-AVC Encoders. In: Picture Coding Symposium (PCS), San Jose (2013)
22. Sullivan, G.J., Wiegand, T.: Video Compression—From Concepts to the H.264/AVC Standard. Proceedings of the IEEE 93(1), 18–31 (2005)
23. Stockhammer, T., Hannuksela, M.M.: H.264/AVC video for wireless transmission. IEEE Wireless Communications 12(4), 6–13 (2005)
24. Ellul, G.-A., Debono, C.J.: An Evaluation of Multi-view Video Transmission Performance on Long Term Evolution Networks. In: International Conference on Telecommunications and Multimedia (TEMU), Heraklion (2014)

Partial Differential Equation (PDE) Based Image Smoothing System for Digital Radiographic Image

Suhaila Abd Halim[1], Arsmah Ibrahim[1], and Yupiter HP Manurung[2]

[1] Center of Mathematical Studies, Faculty of Computer and Mathematical Sciences,
Universiti Teknologi MARA, 40450 Shah Alam, Selangor, Malaysia
[2] Faculty of Mechanical Engineering, Universiti Teknologi MARA,
40450 Shah Alam, Selangor, Malaysia
{Suhaila.Abd-Halim,Arsmah.Ibrahim,
Yupiter.HP-Manurung}@tmsk.uitm.edu.my

Abstract. Over the last few decades, partial differential equations (PDEs) have become one of the significant mathematical methods that are widely used in the current image processing area. One of its common applications is in image smoothing which is an essential preliminary step in image processing. Smoothing is necessary because it affects the result of further processes in image processing. In this project, a system based on second-order PDE and fourth-order PDE models are developed and implemented in digital radiographic image that contain welding defects. The results obtained from these models show better image quality as compared to conventional filters, such as median filter and Gaussian filter. The system is beneficial in assisting radiographic inspectors to produce a better evaluation and analysis on defects in welding images. In addition, non-destructive testing consultants from industries and academician from universities can also utilize this system for training and research purposes.

Keywords: partial differential equation, image processing, image smoothing, second-order PDE, fourth-order PDE.

1   Introduction

Radiography is a non-destructive testing (NDT) methods used by modern industries that can be divided into conventional radiography and digital radiography (DR). The conventional radiography uses film while DR uses a detector to store digitized images. The DR presents several advantages such as immediate availability of the image, interactive parameter setting on devices, digital operation on the images and sharing of the images [1]. The 'real time' application of DR makes the system easier to handle with the appropriate digital image processing and advancement of technology. By using image processing, the quality of digital images can be improved and enhanced. There are numerous software programs such as Quick MTF, I See, ImageJ and Imatest which have served to improve and evaluate image quality [2].

During image acquisition process, images are often corrupted by noise due to environmental condition such as light levels and sensor temperature. The existence of

© Springer Science+Business Media Singapore 2015
M.W. Berry et al. (Eds.): SCDS 2015, CCIS 545, pp. 198–207, 2015.
DOI: 10.1007/978-981-287-936-3_19

noise cause the images can't be used directly and require some enhancement for further process. Generally, image smoothing is applied to filter the existence of noise. Noise filtering is done by replacing the pixel intensity to remove relevant noises while the size of the filter controls the degree of smoothing. The area of image smoothing is still an open research as the quality is a subjective matter. The existing filters such as median, Gaussian and Wiener are able to remove noise, but produced blurring effect on image because the image edges are over smooth [3].

Median filter is a well-known filter because it has capabilities of noise reduction with less blurring effect [4]. The median filter replaces each of the pixel intensity with the median value in the region from a neighborhood of nxn mask. On the other hand, the Gaussian filter uses nxn mask with the weights that are computed according to the Gaussian function:

$$g(i, j) = \frac{1}{2\pi\sigma^2} e^{\frac{i^2 + j^2}{2\sigma^2}}. \tag{1}$$

where $g(i, j)$ is the denoised image in i^{th} row and j^{th} column, σ is the standard deviation that controls the degree of smoothing.

Due to the drawback of conventional filters, the nonlinear partial differential equation (PDE) for image processing, including image smoothing has risen since the late 1980s [5]. The PDE is applied as image smoothing to remove noise while preserving the edge. This produced better quality image as compared to the conventional filters.

Chhabra, Dua and Malhotra [6] demonstrated a comparative analysis of several filtering techniques applied on CT scan images and they concluded that anisotropic diffusion is a promising technique compared to median filter, wavelet decomposition, wave atom decomposition, wiener filter, anisotropic diffusion and NL-means filtering.

In the past few years, the evolution of second-order PDE and fourth order PDE have gain lots of attentions for image smoothing. Second-order PDE has been justified as a good noise removal and edge preservation for digital images [7][8] and fourth-order PDE has been successfully attempted as image smoothing on digital images to reduce the blocky effect [9].

Second-order PDE that was initiated by Perona and Malik [10] has combined the theory of diffusion in image processing. Wee, Chai and Supriyanto [11] had extended the conventional four spreading diffusion directions to eight spreading diffusion direction on ultrasound image with speckle noise. The extended version of second-order PDE is able to retain the edge of image feature. Shanmugam and RSD [12] proposed condensed anisotropic diffusion that was successfully applied on foetus image and real US pediatric images. The method aims to enhance the visual interpretation of radiologist. The condensed anisotropic diffusion consists of two terms that are coherent diffusion term and regular term that is numerically implemented using finite difference approach.

The implementation of fourth-order PDE is to avoid the blocky effects of second-order PDE [9]. Hajiaboli proposed a modification of You and Kaveh [13] that produces fast convergent filter with the ability of good edge preservation [14]. Kumar, Kaushik, Anuradha and Saxena [15] proposed a new hybrid model of line based edge detector and fourth-order PDE with median filter to detect the presence of breast cancer on mammogram image in which the proposed model give better peak signal to noise ratio (PSNR) value.

This paper describes the development of the image smoothing system using PDE-based models that based on second-order PDE by Perona Malik [10] and fourth-order PDE by You and Kaveh [13]. Then, the performances of the models are evaluated and compared with the conventional filters.

2   Materials and Methods

In designing the system, the numerical implementation of second-order PDE and fourth-order PDE using finite difference approach (FDA) are applied. The results from these two models are compared to the conventional filters of median and Gaussian. Figure 1 illustrates the various processes embedded in the image smoothing system.

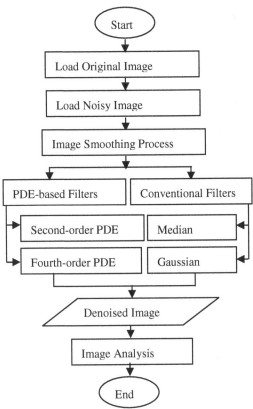

Fig. 1. Embedded Processes in Image Smoothing System

The detail explanation on image smoothing and the development of the system using MATLAB GUI are elaborated in the next section.

2.1   Image Smoothing

In this paper, the PDE-based models applied are (2) represents the second-order PDE by Perona and Malik [10] and (3) that represents the fourth-order PDE by You and Kaveh [13].

$$\frac{\partial I}{\partial t} = \Delta \cdot \left(\alpha \left(\| \Delta I \| \right) \Delta I \right). \tag{2}$$

$$\frac{\partial I}{\partial t} = -\Delta^2 \left(\alpha \left(\left| \Delta^2 I \right| \right) \Delta^2 I \right). \tag{3}$$

where I is an image function of (x, y, t). The $\Delta \cdot$, Δ and Δ^2 are the image divergence, gradient and Laplacian respectively. The α parameter is the diffusivity coefficient defined by Perona and Malik [6] defined as in (4).

$$\alpha(s) = \frac{1}{1 + \left(\dfrac{s}{K} \right)^2}. \tag{4}$$

where K is the contrast parameter.

Both models are solved computationally using FDA. The FDA equation is developed using forward in time and central in space (FTCS) by discretizing (2) and (3) with step sizes of Δx and Δy for the second-order PDE and the step sizes of $2\Delta x$ and $2\Delta y$ for the fourth-order PDE. The discretization for both models produces (5) and (6) respectively.

$$I_{i,j}^{n+1} = I_{i,j}^{n} + \mu \alpha_1 \left(I_{i+1,j}^{n} - 2I_{i,j}^{n} + I_{i-1,j}^{n} \right) + \mu \alpha_1 \left(I_{i,j+1}^{n} - 2I_{i,j}^{n} + I_{i,j-1}^{n} \right). \tag{5}$$

$$
\begin{aligned}
I_{i,j}^{n+1} = I_{i,j}^{n} &- \mu \alpha_2 \left(I_{i-2,j}^{n} - 4I_{i-1,j}^{n} + 6I_{i,j}^{n} - 4I_{i+1,j}^{n} + I_{i+2,j}^{n} \right) \\
&- \mu \alpha_2 \left(I_{i,j-2}^{n} - 4I_{i,j-1}^{n} + 6I_{i,j}^{n} - 4I_{i,j+1}^{n} + I_{i,j+2}^{n} \right).
\end{aligned} \tag{6}
$$

where $\alpha_1 = \dfrac{\Delta t}{(\Delta x)^2} = \dfrac{\Delta t}{(\Delta y)^2}$ and $\alpha_2 = \dfrac{3\Delta t}{4(\Delta x)^2} = \dfrac{\Delta t}{4(\Delta y)^2}$. Δt (delta_T) is the time step and Δx (delta_x) and Δy (delta_y) are the spatial step sizes.

2.2   System Development

The design of Graphical User Interface (GUI) must ease users to communicate with the system in which users are able to manipulate the system to produce the required output. A good GUI should consider three main factors mainly usability, simplicity and interactivity [16].

Usability relates to the interaction of the user interface (UI) which can be defined as user friendly, intuitive or easy to use. For simplicity, the UI must be simple and designed as easy to learn with high proficiency. An interactive UI should provide the way that users expect the system to do. A good interaction of UI will reduce uncomfortable feeling such as stress, fatigue and frustration of users for further use of the system.

In this paper, a system GUI is designed to display the developed image smoothing algorithm that allows users to perform the smoothing process on the input image. Figure 2 demonstrates the layout design of the image smoothing GUI.

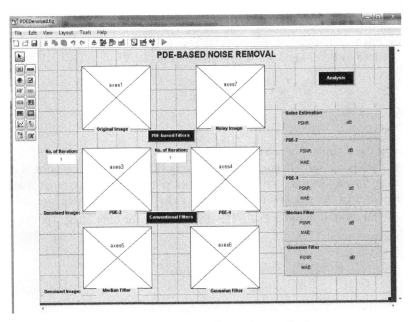

Fig. 2. Image Smoothing System Layout Design

There GUI is designed with six axis graph to display the relevant images where two axes are to display the input images and the other four axes are to display the results of denoised images from the smoothing filters.

In addition, there are three main push buttons required that are PDE-based Filters, Conventional Filters and Analysis. The PDE-based Filters button is developed to execute the process that involved in the second-order and fourth-order PDEs. Before executing each process, users are required to set the number of iterations. The Conventional Filters button is developed to process the median and Gaussian filters. In MATLAB, median and Gaussian filters are processed using:

```
>> B1 = medfilt2(input_image, [3,3]);
>> h = fspecial('gaussian', [3 3]);
>> B2 = imfilter(input_image,h);
```

where B1 and B2 are the processed image for median and Gasussain filters respectively, and the mask used is 3×3 in size.

The Analysis button is created to calculate the performance of both filters (i.e PDE-based and conventional) using full reference image quality metrics that are peak signal to noise ratio (PSNR) and mean absolute error (MAE). The PSNR and MAE are used to measure the noise estimation between the original and the processed image for both models.

The PSNR is the most widely used image quality metric which can be defined as in (7).

$$PSNR\left(I_0, I_{i,j}^{n+1}\right) = 10 \log_{10} \frac{\max\left(I_0\right)^2}{MSE\left(I_0, I_{i,j}^{n+1}\right)}. \tag{7}$$

where I_0 represents the original image and $I_{i,j}^{n+1}$ represents the image produced after the smoothing process. Both I_0 and $I_{i,j}^{n+1}$ are in $A \times B$ size. The $\max(I_0)$ is the maximum possible value of the original image and the MSE is defined as in (8).

$$MSE\left(I_0, I_{i,j}^{n+1}\right) = \frac{1}{AB}\left(\sum_{i=1}^{A}\sum_{j=1}^{B}\left(I_0 - I_{i,j}^{n+1}\right)^2\right). \tag{8}$$

From the calculated PSNR values, higher value indicates a good quality of image and low value shows a less quality of image.

Equation (9) represents the MAE that computes the absolute error between the original image and the processed image by measuring the closeness of the pixels.

$$MAE\left(I_o, I_{i,j}^{n+1}\right) = \frac{1}{AB}\left(\sum_{i=1}^{A}\sum_{j=1}^{B}\left|I_o - I_{i,j}^{n+1}\right|\right). \tag{9}$$

3   Results and Discussion

In this paper, MATLAB GUI development environment (GUIDE) is implemented using a PC workstation with AMD Phenom (tm) II X4 B95 Processor 3.00 GHz running Windows 7 Professional, 32 bit Operating System supported by Matlab R2009a. Table 1 shows the implementation for the developed GUI.

Table 1. The System's Implemention Flow.

Step 1: Read Original Image, I_0
Step 2: Read Noisy Image, I^0
Step 3: Set the number of iterations, n
Step 4: While $n \leq$ number of iterations
Compute (**5**) and (**6**)
end
Step 5: Process the I^0 uses the median and Gaussian filters
Step 6: Calculate *PSNR*, (**7**)
Calculate *MAE*, (**9**)

Figure 3 and Figure 4 depict the load menu and browsing window after users select the load menu. Users are required to load two input images which are the original image, I0 and the noisy image. For the experimental purposes, the noisy image is an image that degraded with some level of Gaussian noise.

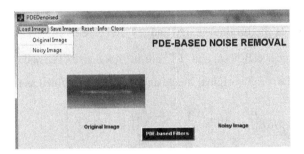

Fig. 3. Layout Design for Loading Image Menu

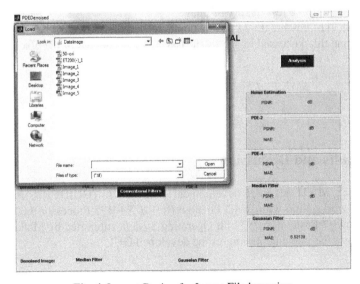

Fig. 4. Layout Design for Image File browsing

After users select a specific input image, users are able to start the smoothing process by pressing the PDE-based Filters button but at first they are required to set the number of iterations. The number of iterations plays an important role that affecting the image quality of the smoothing process for each model. Table 2 depicts the result of the image quality using PSNR and MAE for second-order PDE and fourth-order PDE with different number of iterations for a radiographic image that have been degraded with Gaussian noise.

Table 2. *PSNR* and *MAE* Results with Different Number of Iterations.

Iteration	Smoothing Filters			
	PDE-2		PDE-4	
n	*PSNR(dB)*	*MAE*	*PSNR(dB)*	*MAE*
10	34.8592	2.70677	33.3317	3.24376
50	34.9291	2.61081	35.0177	2.65002
100	33.6517	2.91182	35.144	2.59717
150	32.7029	3.1565	35.094	2.60184
200	31.9661	3.366	35.0145	2.61916

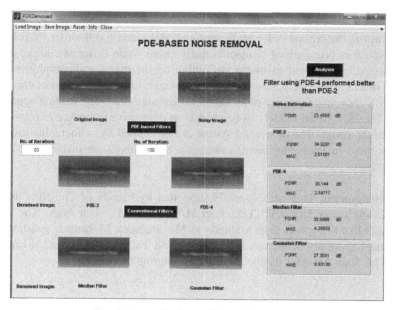

Fig. 5. Layout for Image Smoothing Process

By referring to Table 2, for second-order PDE, n=50 and for fourth-order PDE, n=100 show the best quality results as the highest PSNR indicates the good quality of image. This indicates that the fourth-order PDE produced a better quality image compared to second-order PDE but with higher number of iterations. Then, both n=50 and n=100 are used to display resultant images that required to been analyzed as shown in Figure 5.

It can be seen that, the results of denoised images for second-order PDE, fourth-order PDE, Median and Gaussian filters are hardly to be compare visibly. Hence, the PSNR and MAE for each filter are displayed which indicates the performance for each filter as tabulated in Table 3.

Table 3. The Performance of Image Smoothing Filters using *PSNR* and *MAE*.

Quality Metric	Smoothing Filters			
	Median	Gaussian	PDE-2 *(n=50)*	PDE-4 *(n=100)*
PSNR	30.5889	27.3041	34.9291	35.144
MAE	4.36659	6.53139	2.61081	2.59717

By comparing the image smoothing filters applied in this paper, the quality of radiographic image is improved with the highest PSNR and the lowest MAE using fourth-order PDE.

4   Conclusion

In conclusion, the second-order by Perona and Malik and fourth-order PDE by You and Kaveh that been solved computationally using finite difference approach had been successfully applied as image smoothing on the digital radiography image of welding defect. The developed system aims to demonstrate the implementation of the PDEs with conventional filters of median and Gaussian filters and display appropriate results. The PDE-based model can be used as an alternative technique to improve the quality of the image. The developed system can be applied by radiographic inspectors to assist them in producing better evaluation and analysis on images that contain welding defect.

Acknowledgments. This work is partly supported by Faculty of Computer and Mathematical Sciences, Universiti Teknologi MARA (UiTM), Shah Alam. The authors also would like to express their gratitude to Mr. Shahidan Mohamad, assistant engineer at the Laboratory of Advanced Manufacturing, Faculty of Mechanical Engineering, UiTM, Shah Alam for the technical support during this work.

References

1. Berthel, A., Bonin, T., Cadilhon, S., Chatellier, L., Kaftandjian, V., Honorat, P., et al.: Digital Radiography: Description and User's Guide. In: International Symposium on Digital industrial Radiology and Computed Tomography, Lyon, France (2007)

2. Noorhazleena.: Computed radiography (CR) Signal to Noise Ratio (SNR) Study based on Thickness Changes of Steel Step Wedge. Malaysian Society for Non-Destructive Testing, MSNT (2010)
3. Leavline, E.J., Singh, D.A.A.G.: On Teaching Digital Image Processing with MATLAB. American Journal of Signal Processing 2014 4, 7–15 (2014)
4. Gonzales, R.C., Woods, R.E.: Digital Image Processing. Prentice Hall, New Jersey (2002)
5. Ery, A.C., David, L.D.: Does Median Filtering Truly Preserve Edges Better Than Linear Filtering? The Annals of Statistics 37, 1172–1206 (2009)
6. Chhabra, T., Dua, G., Malhotra, T.: Comparative Analysis of Denoising Methods in CT Images. International Journal of Emerging Trends in Electrical and Electronics 3, 56–59 (2013)
7. Wang, H., Wang, Y., Ren, W.: Image Denoising using Anisotropic Second and Fourth Order Diffusions Based on Gradient Vector Convolution. Journal on ComSIS 9, 1493–1511 (2012)
8. Jing, J., Kehu, Y., Jianjun, G., Wensheng, Y.: Journal of Institute of Electrical and Electronics Engineers, 332–334 (2007)
9. Liu, T., Xiang, Z.: Journal on Mathematical Problems in Engineering 2013, 1–7 (2013)
10. Perona, P., Malik, J.: Scale-Space and Edge Detection using Anisotropic Diffusion. IEEE Transactions on Pattern Analysis and Machine Intelligence 12, 629–639 (1990)
11. Wee, L.K., Chai, H.Y., Supriyanto, E.: Computerized Anisotropic Diffusion of Two Dimensional Ultrasonic Images using Multi-Direction Spreading Approaches. WSEAS Transactions on Biology and Biomedicine 3, 102–112 (2011)
12. Shanmugam, K., RSD, W.: Condensed Anisotropic Diffusion for Speckle Reduction and Enhancement in Ultrasonography. EURASIP Journal on Image and Video Processing 2012, 1–17 (2012)
13. You, Y.L., Kaveh, M.: Fourth-Order Partial Differential Equations for Noise Removal. IEEE Transactions on Image Processing 9, 1723–1730 (2000)
14. Hajiaboli, M.R.: An Anisotropic Fourth-Order Partial Differential Equation for Noise Removal. In: Tai, X.-C., Mørken, K., Lysaker, M., Lie, K.-A. (eds.) SSVM 2009, vol. 5567, pp. 356–367. Springer, Heidelberg (2009)
15. Kumar, A., Kaushik, A.K., Anuradha, S.D.: A New Hybrid Model to Detect the Mammogram Images for Early Breast Cancer. International Journal on Emerging Technologies 1, 28–31 (2010)
16. Hartson, H.R.: Human Computer Interaction: Interdisciplinary Roots and Trends. The Journal of Systems and Software 43, 103–118 (1998)

Main Structure of Handwritten Jawi Sub-word Representation Using Numeric Code

Roslim Mohamad[1], Mazani Manaf[2], Rose Hafsah Abd. Rauf[2],
and Mohammad Faidzul Nasruddin[3]

[1] Department of Computer and Mathematical Sciences, UiTM Kelantan,
18500 Machang, Kelantan, Malaysia
[2] Faculty of Computer and Mathematical Sciences, UiTM Shah Alam, Selangor, Malaysia
[3] Faculty of Information Science and Technology, UKM Bangi, Selangor, Malaysia
rosli027@kelantan.uitm.edu.my,
{mazani,hafsah}@tmsk.uitm.edu.my,
mfn@ukm.edu.my

Abstract. Feature extraction is an important stage in Jawi recognition system because it can influence various aspects that can affect the recognition performance. Statistical feature extraction is strongly influenced by the presence of pixels that make up a word, especially for technique based on zonings and pixel density. Variability in writing style makes the presence of pixels that form the smallest primitive structure in a zone becomes less uniform and this affect the value of pixel density. To overcome this problem, a technique known as numeric code representation to represent the range of the primitive structure tilt in a zone has been proposed. Numeric code is generated by comparing average row and column of smallest primitive structure in each zone. The experimental results show that the numeric code representation is the best method in representing the main structure of the Jawi sub-word image when compared with the other three feature representation techniques. This is because it can generate the highest recognition rate for both classifiers which is used either based on probability or voting.

Keywords: Jawi text recognition, Arabic text recognition, feature extraction, statistical feature. Segmentation-free method.

1   Introduction

Jawi is an old script used in writing Malay language. It was borrowed from Arabic characters by adding additional characters for suitability of use in the language. Jawi script was commonly used in South East Asia where Malay language is spoken by more than 200 million people. However, the use of Jawi script has increasingly been forgotten and replaced with other scripts such roman or script of state official language [12, 15].

Research in the Jawi text recognition is still less compared to the Arabic text or other text that are borrowed from Arabic character. It began in the late 1990s by a

© Springer Science+Business Media Singapore 2015
M.W. Berry et al. (Eds.): SCDS 2015, CCIS 545, pp. 208–217, 2015.
DOI: 10.1007/978-981-287-936-3_20

group of researchers to develop automated digitizing system of Malay Manuscripts at the National University of Malaysia [11]. This work is motivated by the existence of about 15000 ignored Jawi manuscripts in libraries and museums around the world.

Feature extraction is a process to extract a set of features from the image to be classified. Features can be categorized into three types, namely structural, statistical and global transformation feature [7]. Structural feature represent the geometrical and topological characteristics of an image. Statistical features are numerical measures computed over images or regions of images. Meanwhile, global transformation feature represents the abstract or more compact form of image characteristic [2].

The advantages of structural features are that they can tolerate the distortion and variation style of writing. However, these types of features are not easy to extract from the image [6]. The advantages of statistical features are easy to extract from the image. However, they may be misleading due to poor binarization process [7].

The choice of features will affect several aspects of the pattern recognition problem such as accurate representation of word image, learning time, and the required number of sample for training [3]. Good features should have four characteristics namely discrimination, reliability, independence, and small number [4]. However, the main problem in feature extraction is to find the ideal set of features that can be used to make a good and correct classification [9].

As a major factor influencing recognition performance, features play a very important role in handwriting recognition. This has led to the development of a variety of features and feature extraction method for handwriting recognition to increase the performance of system [13]. The complexity of Arabic/Jawi writing style also has led some researchers to extract multiple sources of information from input images [1, 3, 5].

Many Arabic/Jawi characters have the same shape of the main body but only differ in the number and position of dots whether above or below the main body. In this case, dot profile feature become a natural method to differentiate between Arabic/Jawi characters which share the main structure [8].

Variability in writing styles makes feature representation based on calculation of pixel density over the region of image become less effective. So, this paper proposes a technique based on calculation of the pixel average according to their rows and columns in each zone to represent as a numerical code.

2   Sub-word Model

Generally segmentation-free approach based on window is started by implementing implicit segmentation on word or sub-word image to convert the image to a sequence of windows either with uniform or non-uniform width size. Each resulting window will contain the primitive structure of the word or sub-word image. Feature will be extracted from every window to produce the sequence of feature vector to represent the model of a word or sub-word.

This study uses a sub-word model; it is to reduce the number of models that must be developed if compared with the use of the word model. Each sub-word image will be divided into several non-uniform windows where the size of each window is determined by carrying out analysis of the number of black pixels in a column. Each of the resulting windows will contain primitive structure known as the connection structure or core structure as shown in Fig.1.

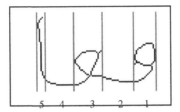

Fig. 1. Windows *1*, *3*, and *5* contain the core structures, whereas windows *2* and *4* contain the connection structures

For the purposes of sub-word model representation, each resulting window will go through the selection process to choose the window that will represent a sub-word model. This situation is different with the usual techniques used in this approach where every resulting window will be selected to represent a model. Selection is made on the basis of the number of core and connection structure produced during the implementation of implicit segmentation. The selection process is very important because it can ensure only the windows that contain the important structures are used to represent the model.

The advantages of this modeling technique are that the primitive structure contained in each window has less variation and only important structures of the sub-word are chosen to represent the model. These factors will give an advantage during the feature extraction and classification stages.

3   Feature Extraction

Two types of features were extracted from each of the resulting windows: statistical (numeric code) from the main structure of sub-word and structural (dot profile). Extraction of dot profile is described in next section.

Extraction of numeric code from main body of sub-word involves two main steps. First, window division, this stage is to divide the selected window into nine zones which is known as smallest window using 3x3 windows as shown in Fig. 2.

f_6	f_3	f_0
f_7	f_4	f_1
f_8	f_5	f_2

Fig. 2. 3x3 windows with nine smallest windows.

The size of the smallest window depends on the width and height of the primitive structure in each of the resulting windows. In some circumstances, a zero smallest windows are used when the conditions of height and width of primitive structure do not fulfill the condition set. Zero smallest window mean the window does not contain any smallest primitive structure and it does not involved any analyzing process on the pixels "1" during the determination of numeric code in the next stage.

Second, numeric code determination; this stage is to represent the smallest primitive structure in each smallest window as a numeric code. Based on the observations of the resulting smallest primitive structure, two categories of the structure were identified, namely the basic and additional structure.

The basic structure is often produced when compared to additional structure. It has straight lines (or nearly straight line) or curve (or slightly curved) structure and has variation of tilt from 0 to 90 degree as shown in Fig. 3, while additional structure has branches or zigzag structure as show in Fig. 4.

Fig. 3. Basic feature of smallest primitive structure.

Fig. 4. Extra feature of smallest primitive structure.

Numeric code determination is started by performing an average calculation of columns (AV_c) and rows (AV_r) of pixels "1" in each of the smallest window. Here is the formula used:

$$AV_c = \frac{\text{pixel "1"}}{\text{column that contain pixel "1"}} \tag{1}$$

$$AV_r = \frac{\text{pixel "1"}}{\text{row that contain pixel "1"}} \tag{2}$$

The average result of rows and columns in each zone will be compared with each other to determine the numeric code that represents the smallest primitive structure. Numeric code 7 is to represent the additional structure, 1 to 5 is to represent the basic structure depending on the difference between the average of rows and columns. Numeric code 0 is to represent the smallest window that does not contain any smallest primitive structure or contain only one pixel "1". Each window is represented by a feature vector fvi = {f0,f1,f2,f3,f4,f5,f6,f7,f8}. However the windows with connection structure will assign fix value of numeric code, fvi = {0,0,5,0,0,5,0,0,5}. This is based on the assumption that all connection structure has straight line shape. The following shows the algorithm of numeric code determination:

```
Begin (numeric code determination)
If (AVᵣ > AV_c)
  {    if (AVᵣ - AV_c) > 0.2
           code = 3
       else if (AVᵣ > 2.2 dan AV_c > 1.8)
           code = 7
       else if AVᵣ > 2.5 and AV_c < 1.1
           code = 1
       else
           code = 2
  }
else if (AVᵣ < AV_c)
  {      if (AV_c -AVᵣ )< 0.2
           code=3
       else if (AV_c > 2.2) && (AVᵣ > 1.8))
           code=7
       else if ((AV_c > 2.5) && (AVᵣ < 1.1))
           code=5
       else
           code = 4
  }
else if (number of pixel "1" equal to 1)
         code=0
else if (number of pixel "1" greater than 1)
         code=3
else
         code=0
end (numerical code determination)
```

The numeric code that represents the basic structure is actually intended to represent the tilt angle of basic structure in a smallest window. Table 1 shows the relationship between numeric code and an estimated range of tilt angle. The tilt angle of smallest primitive structure is measured based on the vertical axis.

Table 1. Numeric code and its estimated range of tilt angle.

Code	Tilt angle
1	$0°$ to $20°$
2	$21°$ to $39°$
3	$40°$ to $50°$
4	$51°$ to $69°$
5	$70°$ to $90°$

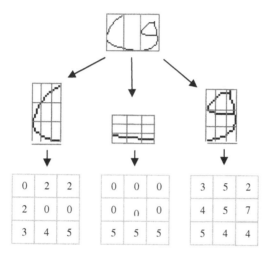

Fig. 5. An example of the production of the smallest windows and the numeric code.

Fig. 5 shows an example of the production of the smallest windows and the numerical code. The resulting feature vector sequence based on figure 5 is as follow:

Fv=2,7,4,5,5,4,3,4,5,0,0,5,0,0,5,0,0,5,2,0,5,2,0,4,0,2,3.

The advantages of this approach is that it is not influenced by the size of sub-word as a 3x3 window size changes according to the size of the resulting primitive structure. Second, it can reduce the effect of variability of writing style because each of the smallest primitive structure is represented as a numeric code based on the estimated range of inclination. It is suitable for use especially in handwritten text recognition systems which are to varied and depend on the writing style.

4   Experiment and Data

The experiment carried out was to compare the efficiency of the technique which is numeric code representatives (tCode) in representing the main structure of Jawi sub-word with three other techniques that use the same approach, namely zoning but varied in terms of statistical features extracted.

The first comparison technique (averRC) extracts two features from every smallest window that is an average of columns and rows. While the second technique (densy1) extract just one feature of pixel density. The three techniques that are tCode, averRC, and densy1 use the same zoning technique based on 3x3 windows.

For the third comparison technique (densy2), zoning methods used were adapted from techniques used by [14]. In this technique, each window is divided into 10 zones as shown in figure 6. Each zone will be represented by one pixel density feature. This technique acts as a control technique.

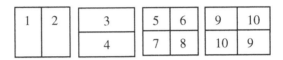

Fig. 6. Zone division in each window

Since this experiment was to determine the efficiency of representative features for the main structure of the Jawi sub-word, so all these three techniques use the same method with the proposed method for representing the dot profile features. Dot profile has been chosen as representative feature because there are lot of main body of Jawi sub-words which has the same main body structure and distinguished by the dot profile only. In Jawi recognition system, the using of dot profile feature can improve the recognition rate up to 10% based on word model [16]. Meanwhile, a study conducted by the authors on sub-word model, it can increase the recognition rate up to 29%. Three profiles of dots were extracted from each sub-word model: the number of dots on a character, the number of dots below a character, and the total number of dots.

Experimental samples (Jawi sub-word handwritten images) are selected from the DbJI database (Database of Jawi Image) from National University of Malaysia. 10 samples of the sub-words were used as training data, and 5 samples of sub-words were used as test data for each lexicon. A total of 1200 samples from 80 lexicons were used.To make the experiment more significant, three sets of training and test samples were prepared and labeled as Set A, Set B, and Set C. Each set of samples was randomly selected from the whole sample to be used as either training or test samples. Figure 7 shows the sample of Jawi sub-words from ten lexicons used in the experiment.

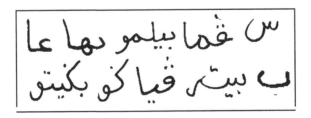

Fig. 7. Experimental sample from ten lexicons.

For classification purposes, this study uses two types of classifier which implements two different classification methods, Multilayer Perceptron (MLP) based on probability and Random Forest (RF) based on the votes. Both of them are from the Waikato Environment for Knowledge Analysis (WEKA) software. The selection of both classifiers is based on the results and consistency for each sample used during experiments to evaluate the performance of the proposed implicit segmentation technique because the tCode extraction technique is only suitable with the implicit segmentation technique.

5   Experiment and Result

The results in tables 2 and 3 show that the tCode technique is more effective in representing the feature of main structure of Jawi sub-word compared with the other three techniques, namely averRC, densy1, and densy2 whether based on probability (MLP) or voting (RF).

The average recognition rate of the tCode technique based on probability is 87.5%, which is 53.4% better than the averRC (34.1%), 0.8% better than the densy1(86.7%), and 9.4% better than the densy2 (78.1%). Meanwhile, The average recognition rate of the tCode technique based on voting is 90.5%, which is 6.4% better than the averRC (84.1%), 3.3% better than the densy1 (87.1%), and 9.6% better than the densy2 (80.9%).

Table 2. Results of experiment using MLP classifier.

	Recognition rate (%) - MLP			
	A	**B**	**C**	**Average**
tCode	89.5	88.3	84.8	87.5
averRC	47	29.5	25.8	34.1
densy1	88.8	87	84.3	86.7
densy2	78	81.5	74.7	78.1

Table 3. Results of experiment using RF classifier.

	Recognition rate (%) - RF			
	A	**B**	**C**	**Average**
tCode	92.3	93.3	86	90.5
averRC	87.8	82.5	82	84.1
densy1	87.5	88.8	85	87.1
densy2	79.3	82.5	81	80.9

The average value of precision and recall in table 4 and 5 also shows that the tCode technique is more effective in representing the feature of main structure of Jawi sub-word compared with the other three techniques, namely averRC, densy1, and densy2.

Table 4. The average value of Precision and recall based on MLP classifier

MLP Classifier				
	tCode	averRC	Densy1	Densy2
Precision	0.894	0.377	0.882	0.803
Recall	0.875	0.341	0.867	0.781

Table 5. The average value of Precision and recall based on RF classifier

RF Classifier				
	tCode	averRC	Densy1	Densy2
Precision	0.919	0.86	0.885	0.833
Recall	0.905	0.841	0.871	0.809

6   Conclusion

Although these four techniques represent the feature from same category, the result of experiment shows that the performance of the feature differs when different techniques of zoning and numerical measurements over region of images were used. tCode technique showed the best recognition rates because the code representation based on estimation of tilt range of the smallest primitive structure are less influenced by the number and pattern of pixel in a smallest window.

References

1. Abed, H.E., Margner, V.: How to Improve Handwriting Recognition System. In: 2009 10th International Conference on Document Analysis and Recognition (ICDAR 2009). IEEE Computer Society (2009)
2. Amara, N.E.B., Bouslama, F.: Classification of Arabic Script Using Multiple Sources of Information: State of the Art and Perspectives. International Journal on Document Analysis and Recognition (IJDAR) 5, 195–212 (2003)
3. Azizi, N., Farah, N., Sellami, M., Ennaji, A.: Using Diversity in Classifier Set Selection for Arabic Handwritten Recognition. In: El Gayar, N., Kittler, J., Roli, F. (eds.) MCS 2010. LNCS, vol. 5997, pp. 235–244. Springer, Heidelberg (2010)
4. Cheung, A., Bennamoun, M., Bergmann, N.W.: An Arabic Optical Character Recognition System Using Recognition-based Segmentation. The Journal of the Pattern Recognition 34, 215–223 (2001)
5. El-Hajj, R., Mokbel, C., Likforman-Sulem, L.: Combining Slanted-Frame Classifiers for Improved HMM-Based Arabic Handwriting Recognition. IEEE Transaction on Pattern Anaysis Machine Intelligence 31(7), 1165–1177 (2009)
6. Haboubi, S., Maddouri, S., Ellouze, N.: Primitive Invariant for Handwritten Arabic Script. International Journal of Computer and Information Science and Engineering 3 (2009)
7. Korsheed, M.S.: Off-Line Arabic Character Recognition – Survey. Pattern Analysis & Application 5, 31–45 (2002)
8. Lorigo, L.M., Govondaraju, V.: Offline Arabic Handwritting Recognition: A Survey. IEEE Transaction on Pattern Analysis Machine Intelligence 28(5), 712–724 (2006)

9. Mazani, M.: Jawi Handwritten Word Recognition System Using Recurrent Bama Neural Network. PhD Thesis, Universiti Kebangsaan Malaysia (2002)
10. Mohammad, F.N., Khairuddin, O., Mohamad, S.Z., Liong, C.Y.: Handwritten Cursive Jawi Character Recognition: A Survey. In: Fifth Int. Conf. on Comp. Graphic. Imaging and Visualization, pp. 247–256 (2008)
11. Mohammad, F.N.: Offline Jawi Handwritten recognition using trace transformation. Thesis Dr. of Philosophy. Universiti Kebangsaan Malaysia (2010)
12. Nor'aiza, A.: Online Non-cursive Jawi Character Recognition system. Thesis Ms., Universiti Malaya (2007)
13. Oh, I.S., Lee, J.S., Suen, C.Y.: Analysis of Class Separation and Combination of Class-Dependent Features for Handwriting Recognition. IEEE Transactions on Pattern Analysis and Machine Intelligence 21(10), 1089–1094 (1999)
14. Ramteke, R.J., Mehrotra, S.C.: Feature Extraction Base on Moment Invariants For Handwriting Recognition. In: Conference on Cybernatics and Intelligent Systems, pp. 1–6 (2006)
15. Razak, Z., Ghani, N.A., Tamil, E.M., Idris, M.Y., Noor, N.M., Salleh, R., Yaacob, M., Yakub, M., Yusoff, Z.B.M.: Off-Line Jawi Handwriting Recognition Using Hamming Classsification. Information Technology Journal 8(7), 971–981 (2009)
16. Remon, R., Khairuddin, O., Mohammad, F.N.: Handwritten Jawi Words Recognition Using Hidden Markov Models. In: International Simposium on InformationTechnology, ITSim 2008, vol. 2, pp. 1–5 (2008)

Part V
Human Machine Interface

Evaluating the Usability of Homestay Websites in Malaysia Using Automated Tools

Wan Abdul Rahim Wan Mohd Isa[*], Muchlisah Md Yusoff, and Dg Asnani Ag Nordin

Faculty of Computer and Mathematical Sciences,
Universiti Teknologi MARA (UiTM), 40450 Shah Alam, Selangor, Malaysia
wrahim2@tmsk.uitm.edu.my,
{muchlisah,asnaninordin}@gmail.com

Abstract. Usability evaluation is an imperative phase in the development of user-centred product design. There are growing numbers of usability research being conducted for different types of websites in Malaysia. However, little research has been done to investigate the usability level of homestay websites in Malaysia. The main objective of this preliminary study is to evaluate the usability of homestay websites in Malaysia by using various automated tools. The study has evaluated 347 homestay websites in Malaysia from the Cari Homestay portal website by using automated tools such as Web Page Analyzer (from Website Optimization) and Dead Link Checker tool. The data were analyzed by using descriptive analysis. The descriptive analysis showed that there are existences of usability issues such as violation of usability guidelines in terms of (i) page size, (ii) broken links and (iii) download speed. Relevant recommendations that can be used by the web developer to improve the website were also being provided. Future work may include series of interviews with real users to share experience on the interaction and perceptions towards the websites.

Keywords: Usability evaluation, automatics tools, homestay websites.

1   Introduction

Usability evaluation is an important phase in the development of user-centred product design. There are growing numbers of usability research being conducted for different types of websites in Malaysia. Examples of usability studies and research in Malaysia include for higher education [1] [2], e-government [3], handicraft [4] and Islamic websites [5] [6]. However, little research has been done to investigate the usability level of homestay websites in Malaysia.

The development of user-centered product design is imperative to satisfy users with ease of use and to keep the producer competitive [7]. Web is a system that is interactive with areas of application but with different levels of usability issues [8]. With growing number of supporting devices, usability evaluation process should be automated [9].

---

[*] Corresponding author.

© Springer Science+Business Media Singapore 2015
M.W. Berry et al. (Eds.): SCDS 2015, CCIS 545, pp. 221–230, 2015.
DOI: 10.1007/978-981-287-936-3_21

2   Usability

Usability of a system can be described as efficient, easy to remember, ease of learning, low error and pleasant [10]. The usability testing or usability measurement used in this study is based on usability traits of (i) download speed, (ii) page size and (iii) broken links of the websites [11]. There is a need to keep page size below 34KB for modem users [11] and the pages accessed should download in less than ten seconds [11] [12]. There are growing numbers of usability research being conducted for different types of websites in Malaysia. Examples of usability studies in Malaysia using various automated tools are as shown in Table 1. Furthermore, Table 2 shows other example of usability research on different websites genre in Malaysia.

Table 1. Example of usability studies in Malaysia using automated tools.

Automated Tool *(Usability traits evaluated)*	Website Genre	Author
Bobby *(Download speed, page size & broken link)*	Islamic	[5]
Website optimization *(Download speed & page size)* & dead-link.com *(Broken link)*	Islamic	[6]
Website optimization *(Download speed & page size)* & Axandra *(Broken link)*	E-Government	[3]
Website optimization *(Download Speed & page size)*, dead-link.com & 1-hitbrokenlinkchecker.com *(Broken link)*	Higher Education	[2]

Table 2. Example of usability research on different website genre in Malaysia.

Website Genre	Usability Research	Author
Handicraft	Usability testing research framework	[4]
Political	Benchmarking framework for assessing usability	[13]
E-Commerce	Usability evaluating using think aloud method	[14]
E-Commerce / Islamic websites	Conceptual model of interaction design process between culture and usability	[15]
Occupational safety & health (OSH)	Usability testing with eye tracking & FCAT analysis	[16]
Online news	Usability measurement	[17]
Higher education	Usability evaluation	[18]
Online tourism	Usability measurement	[19]
Online gift shops	Comparison of usability testing methods	[20]

3   Method

The study had evaluated 347 homestay websites in Malaysia from the Cari Homestay portal website (www.carihomestay.net) by using automated tools such as Web Page Analyzer (from Website Optimization) and Dead Link Checker tool. The data collection process lasted for one month starting from March 2014 to April 2014. The usability information collected in this study are; (i) broken link, (ii) download speed and (iii) uploaded file size of the main page. The two web tools were used in this study in collecting the usability information of the main page. The web tools used are called

Web Page Analyzer (from Website Optimization) available at http://www.website optimization.com/services/analyze/ was used to generate an automatic report on web size and page speed or uploaded time and Dead Link Checker tool, available at http://www.deadlinkchecker.com/ to get the number of broken links.

4   Analysis and Findings

4.1   Overall Results – Usability Issues of Homestay Websites in Malaysia

The usability testing or usability measurement used in this study is based on usability traits such as (i) speed, (ii) page size and (iii) broken links of the websites [11]. Based from guidelines provided; (i) there is a need to keep page size below 34KB for modem users [11], (ii) the pages accessed should download in less than ten seconds [11] [12] and (iii) there should be no existence of broken links [11]. The results are analyzed by using descriptive analysis. In summary, the descriptive results in Table 3 showed that there are existences of usability issues such as violation of usability guidelines in terms of (i) page size, (ii) broken links and (iii) speed.

Table 3. Usability issues of Malaysia homestay website.

Usability Issues	Frequency (/347)	Percentage (%)
Page Size (>34KB)	325	93.7
Broken Link (>=1 link)	230	66.3
Download Speed (14.4k)(>10 seconds)	326	93.9
Download Speed (28.8k) (>10 seconds)	326	93.9
Download Speed (33.6k) (>10 seconds)	326	93.9
Download Speed (56k) (>10 seconds)	325	93.7
Download Speed (128k) (>10 seconds)	316	91.1
Download Speed (1.44Mbps) (>10 seconds)	213	61.4

Table 3 shows the usability issues of homestay websites in Malaysia based on its frequency and percentage. From the data, most of the homestay website in Malaysia (93.7%) had page size of more than 34 KB which may cause the website page to load slower. It shows that most of the homestay website may not follow the usability guidelines. In addition, the speed of the website to load also had severe issues as majority of the main websites had downloaded speed of more than ten seconds. As for the modem speed 14.4k, 28.8k and 33.6k, the numbers of homestay websites with download speed issues are 326 from the total of 347 websites. While for modem speed of 56k, the numbers of websites with download speed issues are 325 websites. There are 316 websites with download speed issues (with modem speed of 128k). The numbers of homestay websites with download speed issues (for the modem speed of 1.44Mbps) are 213 websites. There are 230 websites (66.3%) that had broken links issues.

There are problems associated with usability level of the homestay websites in Malaysia. There are more than 93.9% of the homestay websites takes time more than ten seconds to load their webpage (with 14.4k modem). As the modem speed increases, the download speed of the main page will decrease. With 1.44Mbps modem, there are only 61.4% websites that takes download speed of more than ten seconds to load their main webpage. In addition, the homestay website of Malaysia had other usability issues in terms of the existence of broken link. There are 66.3% of homestay websites in Malaysia with broken links issues. This may be due to the lack of updating of the links available in their website. The existence of broken link may give negative perceptions towards the website as people will think that the page is not up-to-date or indication of poor website maintenance.

4.2   Usability Issues of Homestay Websites Categorized According to Different States and Federal Territories in Malaysia

Table 4 shows the percentage of usability issues for homestay websites categorized according to different states and federal territories in Malaysia. Table 4 shows that there are six states that had 100% page size problems (page size > 34KB) which are (i) Kedah, (ii) Negeri Sembilan, (iii) Pahang, (iv) Penang, (v) Sarawak and (vi) Terengganu. For the broken link problem, (i) Negeri Sembilan and (ii) Terengganu had the highest percentage of problem which is at 86.7%. It is followed with Pahang and Perak (85.7%). Sarawak had the lowest percentage problem of broken link which is only 35.7%. There are problems associated with the download speeds for the main page by using different type of modem speed. For the speed of 14.4k (with download speed of more than ten seconds), there are six states which hold 100% download speed problems which are (i) Kedah, (ii) Negeri Sembilan, (iii) Pahang, (iv) Penang, (v) Sabah and (vi) Terengganu. Melaka hold the lowest percentage which is at 78.6%.

For the speed of 28.8k (with download speed of more than ten seconds), the homestay websites in Kedah, Pahang, Penang, Sabah, Negeri Sembilan and Terengganu hold the 100% for the download speed problem. It is followed by Kelantan by 96.7% for the download speed problem. With modem speed of 56k (page with download speed of more than ten seconds), Kedah, Pahang, Penang, Sabah, Negeri Sembilan and Terengganu recorded as having the page size problem which is 100%. For the modem speed of 128k (with download speed of more than ten seconds), recorded Kedah and Sabah had the highest percentage which is 100%. It is followed with Kelantan, Negeri Sembilan, Penang and Terengganu with the percentage of 93.3%. For the modem of 1.44Mbps speed that had downloaded speed of more than ten seconds, the homestay websites in Kedah and Perlis each recorded the percentages of 81% and 71.4%. It is followed by Sabah (69.2%) with Kelantan and Negeri Sembilan with 66.7%.

Table 4. Percentage of usability issues for homestay websites in Malaysia (States/Federal).

States / Federal Territories*	Page Size (>34KB)	Broken Link (>=1 link)	Download Speed - 14.4k (>10 seconds)	Download Speed - 28.8k (>10 seconds)	Download Speed - 33.6k (>10 seconds)	Download Speed - 56k (>10 seconds)	Download Speed - 128k (>10 seconds)	Download Speed - 1.44mbps (>10 seconds)
Johor (/99)	92/99	61/99	92/99	92/99	92/99	92/99	87/99	56/99
(%)	92.9%	61.6%	92.9%	92.9%	92.9%	92.9%	87.9%	56.6%
Kedah (/31)	31/31	23/31	31/31	31/31	31/31	31/31	31/31	25/31
(%)	100%	74.2%	100%	100%	100%	100%	100%	81%
Kelantan (/30)	28/30	17/30	29/30	29/30	29/30	28/30	28/30	20/30
(%)	93.3%	56.7%	96.7%	96.7%	96.7%	93.3%	93.3%	66.7%
Kuala Lumpur*(/30)	26/30	19/30	26/30	26/30	26/30	26/30	26/30	18/30
(%)	86.7%	63.3%	86.7%	86.7%	86.7%	86.7%	86.7%	60.0%
Melaka (/14)	11/14	10/14	11/14	11/14	11/14	11/14	11/14	9/14
(%)	78.6%	71.4%	78.6%	78.6%	78.6%	78.6%	78.6%	64.3%
Negeri Sembilan (/15)	15/15	13/15	15/15	15/15	15/15	15/15	14/15	10/15
(%)	100%	86.7%	100%	100%	100%	100%	93.3%	66.7%
Pahang (/14)	14/14	12/14	14/14	14/14	14/14	14/14	13/14	8/14
(%)	100%	85.7%	100%	100%	100%	100%	92.9%	57.1%
Penang (/15)	15/15	9/15	15/15	15/15	15/15	15/15	14/15	8/15
(%)	100%	60.0%	100%	100%	100%	100%	93.3%	53.3%
Perak (/14)	13/14	12/14	13/14	13/14	13/14	13/14	13/14	6/14
(%)	92.9%	85.7%	92.9%	92.9%	92.9%	92.9%	92.9%	42.9%
Perlis (/14)	12/14	7/14	13/14	13/14	13/14	13/14	13/14	10/14
(%)	85.7%	50.0%	92.9%	92.9%	92.9%	92.9%	92.9%	71.4%
Putrajaya*(/15)	14/15	9/15	13/15	13/15	13/15	13/15	13/15	6/15
(%)	93.3%	60.0%	86.7%	86.7%	86.7%	86.7%	86.7%	40.0%
Sabah (/13)	12/13	10/13	13/13	13/13	13/13	13/13	13/13	9/13
(%)	92.3%	76.9%	100%	100%	100%	100%	100%	69.2%
Sarawak (/14)	14/14	5/14	13/14	13/14	13/14	13/14	13/14	13/14
(%)	100%	35.7%	92.9%	92.9%	92.9%	92.9%	92.9%	64.3%
Selangor (/14)	13/14	10/14	13/14	13/14	13/14	13/14	13/14	8/14
(%)	92.9%	71.4%	92.9%	92.9%	92.9%	92.9%	92.9%	57.1%
Terengganu (/15)	15/15	13/15	15/15	15/15	15/15	15/15	14/15	7/15
(%)	100%	86.7%	100%	100%	100%	100%	93.3%	46.7%
Total Errors (/347)	**325**	**230**	**326**	**326**	**326**	**325**	**316**	**213**
(%)	93.7%	66.3%	93.9%	93.9%	93.9%	93.7%	91.1%	61.4%

As the data shown from the total usability errors in Table 4, the followings discuss the percentages of homestay websites in different states and federal territories of Malaysia based from the (i) the total usability issue (page size > 34 KB) exists from 325 websites, (ii) the total usability issue (broken link > = 1 link) exists from 230 websites, (iii) the total usability issue (download speed (1.44Mbps) > ten seconds) exists in 213 websites.

Percentages of Homestay Websites in Different States and Federal Territories of Malaysia Based from the Total Usability Issue (Page Size > 34 KB) Exists from 325 Websites

Fig. 1 shows the percentages of homestay websites in different states and federal territories of Malaysia based from the total usability issue (page size > 34KB) exists from 325 websites.

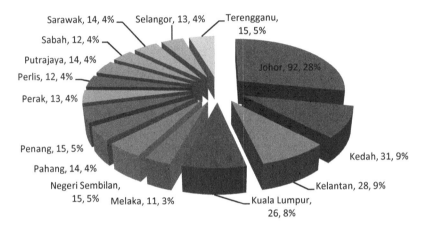

Fig. 1. Percentages of homestay websites in different states and federal territories of Malaysia based from the total usability issue (page size > 34KB) exists from 325 websites.

From Fig. 1, the homestay websites in Johor had 28% of usability issue (page size > 34KB) from the total usability issue (page size > 34KB) exists from 325 websites. The homestay website in Melaka had 3% of usability issue (page size > 34KB) exists from 325 websites.

Percentage of Homestay Websites in Different States and Federal Territories of Malaysia Based from the Total Usability Issue (Broken Link > = 1 Link) Exists from 230 websites.

Fig. 2 shows the percentages of homestay websites in different states and federal territories of Malaysia based from the total usability issue (broken link>=1 link) exists from 230 websites.

Fig. 2 shows that the homestay websites in Johor with usability issue (broken link > = 1 link) of 27% based from the total usability issue (broken link > = 1 link) exists from 230 websites. The homestay websites in Sarawak with usability issue (broken link > = 1 link) of 2% exists from 230 websites.

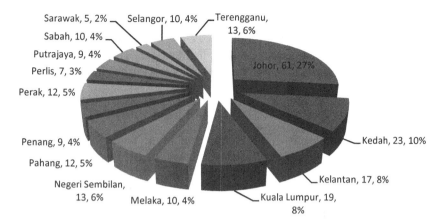

Fig. 2. Percentage of homestay websites in different states and federal territories of Malaysia based from the total usability issue (broken link > = 1 link) exists from 230 websites

Percentages of Homestay Websites in Different States and Federal Territories of Malaysia Based from the Total Usability Issue (Download Speed (1.44Mbps) > Ten Seconds) Exists in 213 Websites

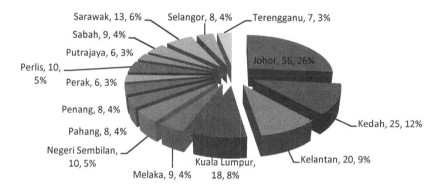

Fig. 3. Percentages of homestay websites in different states and federal territories of Malaysia based from the total usability issue (download speed (1.44Mbps) > ten seconds) exists in 213 websites

Fig. 3 shows that the homestay websites in Johor with 26% of usability issue (download speed (1.44Mbps)>ten seconds) based from the total usability issue (download speed (1.44Mbps)>ten seconds) exists in 213 websites. The homestay websites in Putrajaya, Terengganu and Perak had 3% of usability issue (download speed (1.44Mbps)>ten seconds) based from the total usability issue (download speed (1.44Mbps)>ten seconds) exists in 213 websites.

4.3   Recommendations to Improve the Usability of Homestay Websites

There are some general considerations that a web developer must consider to improve the usability level of homestay websites in Malaysia. The followings are three relevant recommendations to improve the usability level of the homestay websites in Malaysia:

1. There is a need to keep page size below 34KB for modem users [11].

The total sizes of files included in the main page of the website such as images and advertisements must be taken into account. The total size of files in the main website should be 34KB or less. There should be not much of large type of information put in the main page of the website as this may slow down the load of the page.

2. The pages accessed should download in less than ten seconds [11] [12].

It is recommended that the download speed for the main page will not take more than ten seconds to load, as the user might change to another page due to the slower response time. The images used must be in smaller size to improve the download time. It is recommended not to use too many multimedia elements in the main page of the website.

3. There should be no existence of broken links [11].

For broken links, the developer must check all links regularly to ensure that it is accessible. Any changes on link must be updated or the link must be deleted if it is no longer usable.

5   Conclusion

The main objective of this preliminary study is to evaluate the usability of homestay websites in Malaysia by using various automated tools. The study has evaluated 347 homestay websites in Malaysia from the Cari Homestay portal website by using automated tools such as Web Page Analyzer (from Website Optimization) and Dead Link Checker tool. The data were analyzed by using descriptive analysis. The descriptive analysis showed that there are existences of usability issues such as violation of usability guidelines in terms of (i) page size, (ii) broken links and (iii) download speed. Relevant recommendations that can be used by the web developer to improve the website were also being provided. Thus, relevant improvements should be done as to make sure the homestay websites in Malaysia satisfy the usability criteria.

The finding from this study my provide motivations for web developers to give more priority on usability aspect during website development. The findings reported in this study may also alert and create awareness for web developer to give more emphasis on specific usability features which are often being neglected. The limitation of this study is the usability criteria evaluated are only limited to (i) page size, (ii) broken links and (iii) download speed. The other limitation includes usability evaluations were done only on the main page of the homestay websites. The descriptive results were discussed based from the sample gathered from the Cari Homestay portal website and not to be generalized to the populations due to the limitation of the small sample size of the websites. Future work may include series of interviews with real users to share experience on the interaction and perceptions towards the websites.

References

1. Marzanah, A.J., Usman, A.U., Aisha, A.: Assessing the Usability of University Websites from Users' Perspective. Australian Journal of Basic and Applied Sciences 7(10), 98–111 (2013)
2. Abdul Aziz, M., Wan Mohd Isa, W.A.R., Nordin, N.: Assessing the accessibility and usability of Malaysia Higher Education Website. In: 2010 International Conference on User Science and Engineering (i-USEr), pp. 203–208. IEEE (2010)
3. Isa, W.A.R.W.M., Suhami, M.R., Safie, N.I., Semsudin, S.S.: Assessing the Usability and Accessibility of Malaysia E-Government Website. American Journal of Economics and Business Administration 3(1), 40–46 (2011)
4. Isa, W.A.R.W.M., Lokman, A.M., Wahid, E.S.A., Sulaiman, R.: Usability Testing Research Framework: Case of Handicraft Web-Based System. In: The 2nd International Conference on Information and Communication Technology, pp. 199–204. IEEE (2014)
5. Wan Abdul Rahim, W.M.I., Nor Laila, M.N., Shafie, M.: Towards Conceptualization of Islamic User Interface for Islamic Website: An Initial Investigation. In: The International Conference on Information & Communication Technology for the Muslim World, ICT4M (2006)
6. Mehad, S., Isa, W.A.R.W.M., Noor, N.L.M., Husin, M.S.: Muslim User Interface Evaluation Framework (Muslim-UI) for Islamic genre website: A quantitative approach. In: 2010 International Conference on Information and Communication Technology for the Muslim World (ICT4M), pp. H-1–H-6. IEEE (2010)
7. Lee, W., Jung, K., Park, J., Kim, S., Yoon, S., Kim, M., You, H.: Development of a quantitative and comprehensive usability evaluation system based on user needs. In: Human Factors and Ergonomics Society Annual Meeting, pp. 1512–1516 (2009)
8. Yan, P.: Guo. J.: The Research of Web Usability Design. In: The 2nd International Conference on Computer and Automation Engineering (ICCAE), pp. 480–483 (2010)
9. Harms, P., Grabowski, J.: Usage-Based Automatic Detection of Usability Smells. In: Sauer, S., Bogdan, C., Forbrig, P., Bernhaupt, R., Winckler, M. (eds.) HCSE 2014. LNCS, vol. 8742, pp. 217–234. Springer, Heidelberg (2014)
10. Nielson, J.: Usability Engineering. Morgan Kaufmann (1994)
11. Nielson, J.: Designing Web Usability. New Riders Publishing (2000)
12. Chandler, K., Hyatt, K.: Customer-Centered Design: A New Approach to Web Usability. Prentice Hall Professional (2003)
13. Shahizan, H., Norshuhada, S.: Assessing the Usability of Political Web Sites in Malaysia: A Benchmarking Approach. In: Sembok, T.M.T., Zaman, H.B., Chen, H., Urs, S.R., Myaeng, S.-H. (eds.) ICADL 2003. LNCS, vol. 2911, pp. 468–479. Springer, Heidelberg (2003)
14. Majid, R.A., Hashim, M., Jaabar, N.A.A.: An Evaluation on the Usability of E-Commerce Website Using Think Aloud Method. In: Rocha, Á., Correia, A.M., Tan, F., Stroetmann, K. (eds.) New Perspectives in Information Systems and Technologies, Volume 2. AISC, vol. 276, pp. 289–296. Springer, Heidelberg (2014)

15. Wan Mohd Isa, W.A.R., Md Noor, N.L., Mehad, S.: The Information Architecture of E-Commerce: An Experimental Study on User Performance and Preference. In: Papadopoulus G. A., Wojtkowski, W., Wojtkowski, G., Wrycza, S., Zupancic, J. (eds.) Information Systems Development: Towards a Service Provision Society. Springer Science and Business Media (2009)

16. Rashid, S., Soo, S.-T., Sivaji, A., Naeni, H.S., Bahri, S.: Preliminary Usability Testing with Eye Tracking and FCAT Analysis on Occupational Safety and Health Websites. Procedia - Social and Behavioral Sciences 97, 737–744 (2013)

17. Abdullah, R., Wei, K.T.: Usability Measurement of Malaysia Online News Websites. International Journal of Computer Science and Network Security 8(5), 159–165 (2008)

18. Jabar, M.A., Usman, U.A., Sidi, F.: Usability Evaluation of Universities' Websites. International Journal of Information Processing and Management 5(1), 10–17 (2014)

19. Vatankhah, N., Wei, K.T., Letchmunan, S.: Usability Measurement of Malaysia Online Tourism Websites. International Journal of Software Engineering and its Applications 8(12), 1–18 (2014)

20. Goh, K.N., Chen, Y.Y., Lai, F.W., Daud, S.C., Sivaji, A., Soo, S.T.: A Comparison of Usability Testing Methods for an E-Commerce Website: A Case Study on a Malaysia Online Gift Shop. In: 2013 Tenth International Conference on Information Technology: New Generations, pp. 143–150 (2013)

Humanoid-Robot Intervention for Children with Autism: A Conceptual Model on FBM

Azhar Abdul Aziz, Fateen Faiqa Mislan Moghanan, Mudiana Mokhsin, Afiza Ismail, and Anitawati Mohd Lokman

Faculty of Computer and Mathematical Sciences,
Universiti Teknologi MARA, 40450 Shah Alam, Selangor, Malaysia
azhar@tmsk.uitm.edu.my

Abstract. Autism is a lifelong disability that affects children development in terms of social interaction, communication, and imagination. Children with autism often are not able to communicate in a meaningful way with their surroundings and could not relate to the real world. Encompassing humanoid-robot during the therapy session is said as being one of the most beneficial therapies towards these children since autistic children are reported to be keener in engaging in machinery and gadgets. Due to the limited studies in the perspective of the children's emotions and feelings, this study adopts Kansei assessment to investigate the emotions and feelings of the autistic children while engaging with the robot. Kansei assessment was done by the teacher which interpreted the emotional responses given by the autistic children. Two autistic children were involved in the study where both of the subjects are having mild autism. The data were then analyzed and translated to Fogg's Behavioral Model to represent the children's learning motivation. The developed Modified Fogg's Behavioral Model successfully shows the inter-relation between the three components of ability, trigger and motivation for the autistic children while they interact with the humanoid-robot. The final model provides some evidence that despite having limited ability, given the right intervention, the children with autism will exhibit the same level of motivation with normal children.

Keywords: Autism, Ethical module, Kansei, Humanoid-robot intervention, Spiritual module.

1 Introduction

Autism, as described by the National Autism Society of Malaysia (NASOM) is a lifelong developmental disability that blocks children development in learning, language processing and communication emotional as well as social interaction. The term Autism was first introduced by Leo Kanner in 1943 after he identified a group of children that shows an extreme withdrawal behavior from their environment as well as having difficulty in forming a social relationship at a normal age [1]. Children with autism are often being associated with three type of impairment which includes impairment in social interaction, social communication and imagination. The theory on

© Springer Science+Business Media Singapore 2015
M.W. Berry et al. (Eds.): SCDS 2015, CCIS 545, pp. 231–241, 2015.
DOI: 10.1007/978-981-287-936-3_22

the Triad of Impairment for the children with autism was first introduced by Wing and Gould in the year 1979 [2].

Impairment in social interaction has been linked to the children being unable to relate to others or their surroundings, leading them to being isolated and institutionalized [3]. The effort to establish communication with others is a very challenging feat because it is very difficult for autistic children to comprehend what is spoken to them. In addition, the impairment in social interaction also leads to children's inability to make eye contact, to interpret others' feelings as well as voice tones. Autistic children have the tendency to avoid cooperative play resulting them to do their own repetitive activities and eventually eliminate others as well as their surroundings. The inability to communicate with others causes them to avoid getting engaged in any kind of interaction thus preventing these kids from learning many fundamental skills and also hindering their developmental progress. The second impairment which is the impairment in social communication can affect both verbal and non-verbal communication of autistic children where in some extreme cases, speech may be completely absence and children will become totally mute [4]. The third impairment that is the impairment in imagination often causes the children to be less imaginative, which affects their conceptual skills and causing them to be unable to generalize a learnt skill. Despite their limited communication ability, these autistic children are very attracted to technologies such as television and computers [5].

Researchers have started to explore the possibilities of inclusion of technologies, which may provide some glimmer of hopes to these children. Technology intervention for the special education was first explored in the eighties where the computer was used as a medium of transmitting a stand-alone instructional media and graphics to the intended users. The rapid advancement of technology and computer science has allowed a new intervention to the children with autism. Different areas in technology have been experimented upon in order to find the best possible solution for autism therapy. In addition, the ongoing evolution of the technology had led to more new discovery of the computers and machines that is said to be advantageous for the children with autism [3]. The evolvement of technology together with the emergence of artificial intelligence in the past decade had further elevated the chances of a better therapy method and treatment to be explored and implemented. In the recent years, robots have started to make its debut as the mediator between the therapist and the autistic children where these children were said to perform better during their therapy session [3].

Upon recognizing the benefit of using robots to the autistic children, researchers had begun to explore and design various types of robots for the intervention purposes. These led to creation of various types of robots ranging from the simple machine-like robots, soft toys, animals-robot and the latest type is the invention of the robot that resembles a human, which is known as the humanoid-robots. Humanoid is a robot in which its fabrication is akin to human being with all basic structures such as head, hands and legs. Humanoid-robot had been used in the therapy session with the autistic children, in which one of the aims of the intervention is to prepare these children to face the real human world and to make them able to communicate with others in a more meaningful way. These findings have encouraged the exploration of the use of humanoid-robot as a mediator with autistic children. Humanoid robots have a large potential in educating and therapy session with the autistic children where it allows them to detect and learn about different emotion, expression as well as social behavior of human beings [6]. The usage of humanoid-robot is said to provide the children with gradual integration to the real world and later allowing them to lead an independent life of their own [7, 8].

Despite of the promising future brought by the humanoid-robot to the autistic children, studies on the use of humanoid-robot as a therapy tool do not take the children's emotion and perspective into consideration. Previous studies have claimed that despite of the advantages and benefits brought by the robots toward the autistic children, very little has been published on the affective effect of the intervention and on the children's perception towards the robot [5]. The aim of this paper is to present a study that investigates autistic children's feelings and emotion while engaging with the humanoid-robot. With the assistance of the humanoid-robot as a mean to extract the emotion of these children, the research was able to relate the children's emotion to a behavioral model. The result provides an insight to the study related to the emotions of children with autism toward the humanoid-robot interaction, which may later contribute to the best possible therapeutic approach for them.

2   Fog's Behavioral Model

The studies on human behavior have long existed and one of the first model on human behavior was published in 1955 where Ripple first introduced the three-dimensional behavior model encompassing motivation, capacity and opportunity [9]. The emergence of computer sciences and the boom of the Internet technology have accelerated the study on the human behavior [10]. Multi-variance of the social networking site as well as the new applications and software have paved a new way of communication among humans as well as gathering different types of human behavior under one roof. These have become one of the core motivations that expedite the research conducted on the study of human behavior. Along with the increasing number of researches, different researchers have also attempted to develop their own understanding of human behavior which have resulted many different models. One of the more popular models introduced recently is the Fogg's Behavioral Model. Fogg's

Behavioral Model (abbreviated as FBM) was first introduced in 2009 by Dr. B.J Fogg who is also known as the founder of the Persuasive Technology Lab in Stanford University, United States. He has initially introduced the FBM with the intention to help other designers to have a better idea on the persuasive design in technology. He further claimed that FBM could be used in multiple domains ranging from medical to educational research. FBM is used to assist a researcher in the effort to better understand human behavior as well as obtaining clearer insight in persuasive research. Generally, Fogg's Behavioral Model sees human behavior as the product of three factors which is motivation, ability as well as triggers. In order for a particular behavior to occur, an individual must be sufficiently motivated, possesses the ability and appropriately triggered to perform the intended behavior (B=mat). Figure 1 below illustrates the FBM as proposed by Fogg [11].

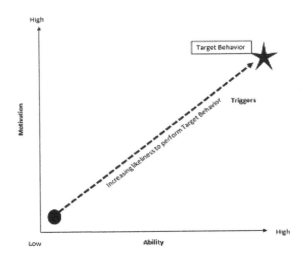

Fig. 1. Fogg's Behavioral Model showing the three dimensions on motivation, ability and triggers [12]

Figure 1 illustrates how the three dimension of motivation, ability and trigger works. Note that FBM does not have any unit on the axis. This is due to the fact that the framework is a conceptual framework and the main purpose of the framework is to define the relationship between the three different dimensions and not the precise value [12, 13]. The model firstly depicts motivation element covers three different domains, which are sensation, anticipation and social cohesion. Secondly, the ability element in the FBM is described as an individual capacity to perform a task. Third element is trigger, defined as an external force that will initiate the subject to perform the target behavior. Trigger for FBM is being separated into 3 sub-elements that are facilitator as trigger, signal as trigger and sparks as trigger. Facilitator as triggers is suitable to be used when the subject is having high motivation but lack in ability where the signal as triggers is suitable to be used when the subjects have high motivation and ability. The third type of triggers, sparks is suitable to be used in the condition where the subjects are having low motivation but is high in ability.

3   Conceptual Framework

3.1   Proposed Modified FBM

FBM was recently introduced but has not been widely explored. To date, FBM is only being used on the study of daily behavior and the founder is encouraging researchers to adapt FBM in various domains including medicine, health as well as in education.

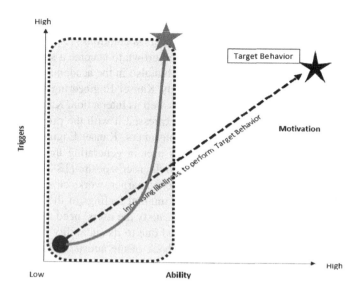

Fig. 2. Proposed Modified Fogg's Behavioral Model (MFBM)

FBM has been said to be able to give researchers an even clearer view on human behavior as well as allowing a researcher to have a better insight on the persuasive research. In this study, based on the original FBM model, the modified version takes the characteristics of the autistic children into consideration. Figure 2 depicts the proposed modified FBM (MFBM) that will be used for the study.

In the proposed modified FBM, the three different factors of human behavior to be investigated which are ability, motivation and trigger, are still maintained as proposed in FBM originally by BJ Fogg's. The founder also added that, researchers were allowed to alter the position of the element within the model to suit the need of the study. In the axis, trigger and motivation is reversed in order to study the degree of how the robot (as trigger) affects the children's learning motivation. The ability axis remains the same but a limitation is being placed to depict the autism children ability that spans from low ability for severe autism to medium ability for mild autism.

4   Methodology

Kansei Engineering is a customer-oriented product development method which takes into consideration the customers' feelings and emotions into the product design element [14, 18-20]. It was first born in Hiroshima in 1970 and rapidly developed as a new branch of ergonomics design [14]. The word "Kansei" originated from the Japanese terms which mean emotions and feelings. [14, 17-19] further explain that the study on Kansei Engineering was first influenced by the research conducted by a German philosopher, Baumgartner which published a study on AESTHETICA [18]. There had been a significant amounts of attempts been made by the researchers to explain and define Kansei but no one single agreed definition were being deduced.

Researchers however had generalize the term Kansei where it can be referred and translated as sensitivity, sensibility, feeling as well as emotions [15], [16].

Kansei Engineering as according to [16] had grown to become a significant discipline not only in the industrial design industry but also in the academic field where the Kansei research to date is not limited to only Kansei Engineering but extended to Kansei information, Kansei communication as well as interaction. Kansei Engineering looks into a consumers' implicit needs and expressed it with the product design that later contributes to the satisfaction of the consumers. Kansei Engineering is a consumer oriented technology that encompasses user in generating the design requirement to develop a single product that match to the user's desire [18-20]. In a nutshell, Kansei, despite of being in any domain mentioned earlier, works closely with the user higher capability of the brain by tackling the implicit feelings of the users and translated these feelings into the product design to satisfy the users' need and desires.

For this study, Kansei evaluation is adopted due to its suitability with the purpose of studying the affective and emotional feedback of the autistic children while interacting with the humanoid-robot NAO. The method begins with identification of the emotional element related to the humanoid robot interaction and also the collection of the humanoid-robot intervention modules for autistic children. Then, the process continues with collecting and identifying the Kansei Word before the collected word was transformed into Kansei checklist. Most of the processes happened at this stage were being conducted qualitatively through an experts' interview. This is due to fewer studies had been performed studying the children's emotion on the humanoid-robot interaction. Then the process resumes with the conduction of the evaluation experiment where the humanoid-robot NAO is being used as the stimulus to interact with the children. Then, the evaluation data are analyzed, interpreted, and later used in deducing the newly Modified Fogg's Behavioral Model.

Table 1. Response coding.

Component	Sub-component	Response
Emotional Responses	Notable Positive Emotion Feedback	
	Notable Negative Emotion Feedback	
Subject Participation	Initial Intervention	
	During Intervention	
	After Intervention	
Learning Motivation	Normal Classroom	
	During Intervention	
	After Intervention	
Ethical and Spiritual Framework	Ethical Framework	
	Spiritual Framework	

Although Kansei evaluation is used as the method to assess the emotion and feeling of autistic children toward the humanoid-robot intervention, note that these children do not possess the ability to interpret and show their feelings directly. Hence, a direct observation was used to observe and study their behavior while engaging with the robot. A video recording session was conducted throughout the intervention session using two internal cameras of the robots and few other external cameras. Additionally, an interview session was conducted with the teachers involved with the intervention session. The interview session was intended to get the children feedback after engaging with the humanoid-robot and also validate the Modified Fogg's Behavioral Model. Result from the observation and interview is coded using Table 1, to provide input to the behavioral model.

Figure 3 shows the interaction module used during the intervention session. Two autistic children with different severity level and one normal child participated in a close experiment setting.

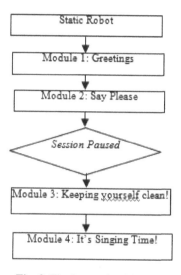

Fig. 3. The Interaction Module

5   Result and Discussion

The result shows that, of all the four modules, the fourth module on singing and dancing were the one that marked the highest score for all the positive emotions in the checklist. According to the teacher, this was due to the fact that children's emotion and feeling were highly triggered with the robot singing and making hand gestures when it dances. To support the Kansei analysis, an interview session was performed with the teachers. The main intention of this interview is to gain the teacher's opinion and view on the data tabulated from the Kansei checklist and how it can relate to the children's learning motivation. Additionally, the study attempts to provide prove in the modified FBM, despite limitation of the autistic children in their ability, given a certain amount of trigger, their motivation can be evoked and perhaps similar to those of a normal children.

The interview session started with getting the teacher's view on the use of humanoid-robot with the autistic children where the teachers mentioned that using robot as the teaching aid did spark a different response from the children. The children, including the non-verbal subject (Mild_2) were showing a different side of her while engaging with the robot.

Upon being asked about the difference in learning motivation, the teacher agreed that comparing to the normal classroom, both subject Mild_1 and Mild_2 were more eager and attracted to learn with the robot. She then further added that ever since the robot intervention being conducted, subject Mild_1 kept on asking about the robot and when he would be able to learn with the robot again. However, for subject Mild_2, her inability to verbally express her needs hindered the teacher from evaluating whether she was motivated to learn with the robot again or not. But post-robot intervention, upon hearing the word robot, or whenever the teachers mention about her previous engagement with the robot, subject Mild_2 would smile and become happy. The interviewer then asked the teacher on any observable changes in the two subjects after the intervention session. The changes obtained here were specifically referred to the children' focus before, during and after the robot-intervention session.

However, the teacher mentioned that it was hard for them to do the comparison. This is due to the reason that both subject Mild_1 and Mild_2 were among the students who love to attend their everyday normal classes. Above that, robot-intervention was being conducted only once and with a limited time given for the children to interact with the robot. Hence, to compare the children' focus and concentration during the robot-intervention and after the robot-intervention was being performed is rather difficult. The teacher then added that comparison is possible to be made between the children focus and concentration before the robot-intervention and during the intervention session. Subject Mild_2 in particular was having quite a severe concentration issue where she can hardly be told to do something or to sit still for more than a few seconds. However, while interacting with the robot, she could "magically" sat in front of the robot for the entire intervention session which took around 10 minutes. In addition to that, her focus was set to the robot where the video footage showed that her eyes followed every movement that the robot made and she did not even turn her head away every time the robot was in action.

Based on the assessment being done after the intervention session, it can be concluded that the children with autism, despite of their limitation in ability, the right triggers will evoke their motivation to perform the preset target behavior. However, due to the limitation in the number of subjects, the study could not generalize whether the same amount of trigger would bring the same result for autistic children with different severity levels. Additionally, the limitation of the number of subject had somehow affected the contour line on the final version of the MFBM. As the data are being analyzed and interpreted, the research had deduced on the following result for the Modified Fogg's Behavioral Model. The following figure shows the plotting on the element ability for the subject:

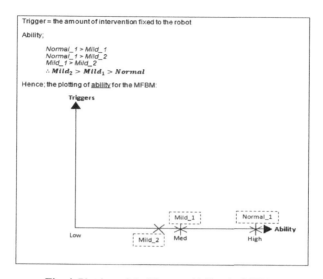

Fig. 4. Plotting of the Element Ability for MFBM

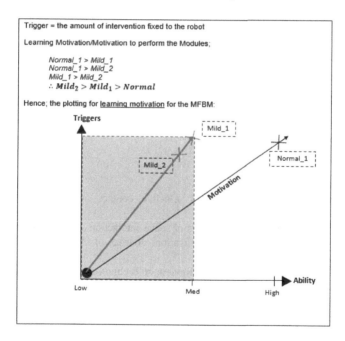

Fig. 5. The Newly Modified Fogg's Behavioral Model

The result of the interview session shows the teachers agreed that not only emotions of the children become visible during the interaction session, but also their learning motivation, which is different from their normal everyday class. The autism subjects showed different sides of behavior during the humanoid-robot intervention and they engaged better compared to when learning with a human teacher.

After the execution of the experiment as well as the evaluation process, the final result for the newly Modified Fogg's Behavioral Model (as is in figure 4) is rather different compared to the earlier proposed model. This is mainly due to the model is being developed based on only two autistic subjects. Relatively, the relationship between the three elements of the model for the children with autism is successfully shown. The line contour depicts by the red line shows that in spite of being limited in their ability, given the right trigger, the children with autism can have equal amount of motivation to perform the targeted behavior.

6   Conclusion and Future Works

Despite of the early concern on possibility in extracting autistic children emotion, the study had successfully discovered that it can be obtained with the use of Kansei assessment. The result as presented in this paper provides evidence that despite being referred to as emotionless, through the use of the right approach, these children' emotions are identifiable. The new MFBM, on other hand had shown that even with their limited ability, given the right triggers, these children can still have the same level of learning motivation as much as normal children. Further study is suggested to include a bigger population size so that the plotting of the MFBM will be more accurate and the interrelationships among the three components can be seen more clearly. Future study should also consider subjects of more diverse severity level so that better insights on the level of motivation over different ability of the children can be identified.

Acknowledgement. The research is supported by the Ministry of Education, Malaysia, under the FRGS grant scheme [Project Code: 600-RMI/FRGS 5/3 (32/2014)].

References

1. Kanner, L.: Autistic disturbances of affective contact. Actapaedopsychiatrica 35(4), 100–136 (1967)
2. Wing, L., Gould, J.: Severe impairments of social interaction and associated abnormalities in children: Epidemiology and classification. Journal of Autism and Developmental Disorders 9(1), 11–29 (1979)
3. Woodward, J., Reith, H.: A Historical Review of Technology Research in Special Education. Review of Educational Research 67(4), 503–536 (1997)
4. Robins, B., Dautenhahn, K., TeBoekhorst, R., Billard, A.: Robotic assistants in therapy and education of children with autism: can a small humanoid robot help encourage social interaction skills? Universal Access in the Information Society 4(2), 105–120 (2005)
5. Cabibihan, J.J., Javed, H., AngJr, M., Aljunied, S.M.: Why robots? A survey on the roles and benefits of social robots in the therapy of children with autism. International Journal of Social Robotics 5(4), 593–618 (2013)
6. Conn, K., Liu, C., Sarkar, N., Stone, W., Warren, Z.: Affect-sensitive assistive intervention technologies for children with autism: An individual-specific approach. In: The 17th IEEE International Symposium on Robot and Human Interactive Communication (ROMAN), pp. 442–447 (2008)

7. Hashim, R., Sh. Ahmad, S., Ismail, Z., Fikry, A.: Ethics in Humanoid Robot Intervention in Social Skill Augmentation of Brain-Impaired Children. In: IEEE Symposium on Business, Engineering and Industrial Application, pp. 51–55 (2014)
8. Robins, B., Dautenhahn, K., Dubowski, J.: Investigating Autistic children's attitudes towards strangers with the theatrical robot-A new experimental paradigm in human-robot interaction studies. In: 13th IEEE International Workshop on Robot and Human Interactive Communication (ROMAN), pp. 557–562 (2004)
9. Hashim, R., Mahamood, S.F.: Humanoid Robots for Skill Augmentation of Gifted Children: Teachers' Perceptions and Islamic Implications. Procedia Computer Science 42, 345–350 (2014)
10. Ripple, L.: Motivation, capacity, and opportunity as related to the use of casework service: Theoretical base and plan of study. The Social Service Review, 172–193 (1955)
11. Song, Y., Zhang, C., Wu, M.: The study of human behavior dynamics based on blogosphere. Web Information Systems and Mining (WISM) 1, 87–91 (2010)
12. Fogg, B.J., Hreha, J.: Behavior Wizard: A Method for Matching Target Behavior with Solutions. In: Ploug, T., Hasle, P., Oinas-Kukkonen, H. (eds.) PERSUASIVE 2010. LNCS, vol. 6137, pp. 117–131. Springer, Heidelberg (2010)
13. Fogg, B.J.: A Behavior Model for Persuasive Design. In: PERSUASIVE 2009 (2009)
14. Nagamachi, M., Lokman, A.M.: Kansei Innovation: Practical Design Applications for Product and Service Development. Taylor & Francis Group, CRC Press, Florida (2015)
15. Lee, S., Harada, A., Stappers, P.J.: Pleasure with Products: Design Based on Kansei, Pleasure with Products: Beyond Usability, pp. 219-229 (2002)
16. Harada, A.: On the Definition of Kansei. In: Modelling the Evaluation Structure of Kansei Conference, vol. 2 (1998)
17. Levy, P.: Beyond Kansei Engineering: The Emancipation of Kansei Design. International Journal of Design 7(2), 83–94 (2013)
18. Lokman, A.M.: Emotional User Experience in Web Design: The Kansei Engineering Approach, Doctoral Thesis, UniversitiTeknologi MARA (2009)
19. Lokman, A.M., Nagamachi, M.: Kansei Engineering: A Beginners Perspective, University Pub. Centre (UPENA) (2010)
20. Lokman, A.M., Haron, M.B.C., Abidin, S.Z.Z., Khalid, N.E.A., Ishihara, S.: Prelude to Natphoric Kansei Engineering Framework. Journal of Software Engineering and Applications 6, 638–644 (2013)

Cross-cultural Kansei Measurement

Anitawati Mohd Lokman and Mohd Khairul Ikhwan Zolkefley

Faculty of Computer and Mathematical Sciences,
Universiti Teknologi MARA (UiTM), 40450 Shah Alam, Selangor, Malaysia
anita@tmsk.uitm.edu.my

Abstract. Kansei Engineering (KE) enable designers in making decisions and focus on the design elements which make the product better fit to human feelings by discovering relationships between the customers' feelings and the product features. However, using paper based checklist in evaluating the experiments lead to different results in different cultural races and demographical background and also limits the potential in obtaining desired results. This research attempts to fill in the gap by providing Web-based Kansei Measurement System and test it across culture to see whether it produce similar results. A comparative study of Kansei by two cultural background subjects and two measurement mechanisms, which are web-based and paper based Kansei checklist have been conducted. The resulted Kansei structure shows encouraging evidence that Web-based Kansei Measurement System could be used as cross-cultural Kansei measurement mechanism. The findings could benefit researchers and designers in the effort to improve the process of Kansei measurement to get the desired results.

Keywords: Affect, Cross-culture, Emotion, KE.

1   Introduction

E-commerce web sites are becoming important to organizations where customers are not only bound to demographic cultural boundaries. The emergence of the web as a communication medium is very significant in information sharing collaborative work platform and social networks [1]. Currently in KE, the main method in evaluating the Kansei is by using paper based verbal self-reporting tool, which in the end provide results that are limited by the uniqueness of human Kansei in different cultural races and demographical background [2]. As a consequence, with the current situation, the measurement experiments will take a long time to complete and the measurement experiments can only be done in local. With the wide use of the Internet, users from all over the world could be able to access any pages available in the World Wide Web. Thus this could create a cross cultural clash between the users when accessing a site. Therefore when designing a web site, it is important to understand cultural differences [3].

According to [4], there are two approaches to distinguish the cultural differences in web design, and they are culturalization and cultural representation. Culturalization is the term used to describe a framework that assumes improving the design of the

© Springer Science+Business Media Singapore 2015
M.W. Berry et al. (Eds.): SCDS 2015, CCIS 545, pp. 242–251, 2015.
DOI: 10.1007/978-981-287-936-3_23

websites is possible by evaluating culture differences which means characteristic of each target culture is taken into account in different versions of an application that is being built. The cultural representation on framework, on the other hand is the commonly used to build application shared by users from different cultures. In this framework the basis of cultural differences is considered to reside in the representations used in the application and the meaning conveyed by these representations.

Web design become more and more crucial in e-commerce web sites as consumers from different cultures had different attitudes, preferences, and values, and remained reluctant to buy foreign products even after considerable exposure to globalization [5]. Consumer behavior in e-commerce situations can also be affected by differences in national culture [6]. The nature of e-commerce of being available across borders brought about the importance to fulfil the preference of cross-cultural needs and expectations. The literature has also shown that understanding cross-cultural needs and preferences is crucial to win cross-culture's consumer's heart and mind.

This study aims to investigate the possibility of integrating web-based Kansei measurement, a Web-based Kansei measurement tool which are developed based on the internet technology. In order to achieve this, empirical investigation on how Web-based Kansei measurement tool works and the comparison of resulted Kansei structure when measuring Kansei with paper-based verbal self-reporting instrument in KE are performed. The domain specimen for the comparative study is website User Interface Design (UID) and specimens were selected based on significant visual design differences that possibly elicit different emotional responses.

2   Research Background

This section will provide literature analysis on the topic addressed in this research.

2.1   Emotion and Culture

In determining the definition of emotion, there is a need to separate emotions from phenomena that are not emotions [7]. As suggested by [8], emotions are best treated as a multifaceted phenomenon that consists of behavioral reactions' components, expressive reactions, physiological reactions and subjective feelings. [7] also posited that the cause of emotions is about what is happening between the stimulus and the emotion or between the stimulus and the consequent emotion episode.

According to [9], emotions exist in all cultures, regardless of the presence or absence of any linguistic notions that correspond to a particular emotion. The cross-cultural perspective in emotion research has a long history and the interest in the relations between culture and emotion began in cultural anthropology where theorist studies the cultural relativity of emotion and the powerful influence of cultural factors on human behaviors [10][11].

According to [12], there are four elements of culture. First, culture is not a distinctiveness of individuals but it is a group of individuals who share general values, beliefs, ideas etc. that include family, occupational, regional or national groups. Second, culture is learned, meaning that culture can be learned once the individuals become member of the culture. Third, culture has historical value which a specific nation's culture develops over time and is part of the nation's history, its demographics and economic aspect, its geographical and ecological environment. Lastly, culture has different layers. This is supported by [13] where he distinguishes four different layers of culture namely symbols, heroes, rituals and values. [13] defines culture as "-the collective programming of the mind that distinguishes and differentiates between member of one group and category of people from another".

Hofstede's Cultural Dimensions Model was used to measure the differences in national culture [14]. [15] quoted "even though, his work has been criticize, his four dimensions of national culture do present a systematic framework for identifying differences in national cultures and embodies the biggest study attempting to categorize nations based on broad value differences". Hofstede's national culture dimensions model is based on distinctions in values and beliefs towards work goals which have implications for business by providing a clear relationship between national and business cultures. He also identifies four dimensions that differentiate different national cultures namely Power Distance, Uncertainty Avoidance, Individualism/Collectivism, and Masculinity/Femininity.

The ability to measure user emotions in computer science has become important for designing intelligent interface, so that computer would able to establish believable interaction or could alter its internal behavior based on the user's emotions [16]. According [8] and [17], "in order to measure emotions, one must distinguish the characteristic of emotion, but as there are various traditions that hold different views on how to defining, studying and explaining emotions, makes measuring a person's emotional state as one of the most difficult problems in affective science". Therefore any tools that said to measure emotions are in fact measuring some of the parts in emotions and in the current common evaluation methods of emotions, the evaluation seeks to a particular kind of emotion [18].

2.2 Product Emotions

The emotional power of products has never been doubted, the emotional attraction that a product poses could become the differential advantage in the marketplace as products are now often similar with respect to technical characteristic, quality and price [19]. Furthermore, a good human-centered design of a product, which has pleasant and pleasurable aspect would makes users more tolerant of any difficulties they encounter while using the product [20].

A product that elicit good feelings to the consumer would have all kinds of likeable effects such as could be talked about in a positive way, contribute to brand image and sometimes are forgiven for design imperfections [21]. In terms of the consumers, owning and using products have an influence on one's identity; it could affect an individual's self-perception and how the individual is perceived by others.

KE become a popular tool to measure product emotion and engineer them into product design [22]. [23] defined the term Kansei as implying psychological feeling and needs in mind. [23] argues that Kansei is the term used in Japanese to describe or express one's impression towards artefact, situation and the surrounding. Engineering on the other hand is defined as 'the application of science and mathematics by which the properties of matter and the sources of energy in nature are made useful to people; the design and manufacture of complex products' [24]. According to [2], KE is designed to capture subjective consumer insight, synthesize them with the actual product design element which is to map what Kansei is associated to which element so that the new product design embeds the consumer insight. It also could be referred as "emotional design" or "sensory engineering" aims to translate consumer psychological feeling about a product into perceptual design element, allowing design and evaluation of products before lunching them to the market [25]. Since Nagamachi first introduce KE, it had been successfully used in incorporate emotional aspect into product design and is well accepted as industrial design method in Japan and Korea but is better known as emotional design in Europe [23].

2.3   Website UID and Emotion

Websites often provide the first impression of an organization and are crucial to the organization [26]. When a user opens a web site, the first impression is probably made in a few seconds, and the user will either stay or move on to the next site on the basis of many factors. The website page aesthetics is one of the many factors in determining the success of the website [26]. A user's perception of a Web site can evoke a wide range of emotions and attitudes and these emotions and perceptions impact the user's attitude towards the Web site's content, advertised products, company, credibility and site usability [27]. According to [27] the primary elements that Web designers use to communicate with users include home page length, graphics, links, text, and animation [28-31, 32]. These elements are used as criteria for specimen selection in this research.

3   Research Methods

Fig. 1 shows Kansei Design Model, which was developed to provide a systematic approach to the implementation of KE in designing Kansei product [31]. The model is a basis for many KE implementations to discover Kansei concept and design requirements for new innovative Kansei product. The research adopts this model, and performs important steps in L1, L2 and L3 to measure and analyze Kansei. The steps that were implemented in this research are identification of specimen from existing product, development of Kansei measurement checklist (both traditional tool and web-based tool), Kansei measurement, and analysis.

The data collected for this research was done using two methods. The first method was by using Web-based Kansei Measurement System. The data collected in Web-based Kansei Measurement System was automatically compiled in a text formatted document and saved using Microsoft Excel application. The data was downloaded to a personal computer as a Microsoft Excel document for analysis purposes.

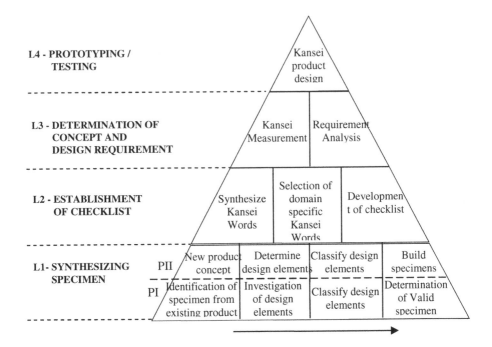

Fig. 1. Kansei Design Model [31].

The second data collection method was based on traditional paper based Kansei checklist. Each participant was given 10 checklists, one checklist for each specimen in the evaluation of their emotional responses towards the Website UID. All checklists were collected after the evaluation is completed, and the data were then entered manually into Microsoft Excel document for analysis purposes.

For the evaluation process, respondents from two different ethnic groups were recruited to obtain required data in order to analyze the cross cultural aspect of Kansei measurement. The two ethnic groups, labelled as Ethnic1 and Ethnic2, were selected based on convenience sampling for comparative analysis purposes. These two ethnics were chosen because they are the largest ethnics in the country. Also, these two ethnics were selected because they have easy access to technology thus suitable for what this research is meant to. Multivariate analysis was then performed to the collected data to analyze the similarities and differences of Kansei structure by both cultures and evaluation methods.

4   Result and Discussions

This section discusses the comparison of the structure of Kansei that formed from evaluation result by two ethnic groups, Ethnic1 and Ethnic2, using paper based Kansei measurement.

Analysis of the structure of Kansei responses and its relations with specimen were observed using Principal Component Analysis (PCA). As evident in Fig. 2 and Fig. 3, the structure of Kansei those forms from both measurement methods are similar. It can be observed that Hope and Boredom are at the opposite space of both PCA plots. At far left of the plots locate Dissatisfaction, Sad and Shame, and at most right of the plot resides Satisfaction, Joy, Fascination and Desire. The distributions of their relations to specimens are also similar from both plots. The result from these plots suggests that the use of Web-based Kansei Measurement is successful and produced desired result. It also indicates that other than simplifying the measurement process and control, the Web-based Kansei Measurement System could produce optimum result.

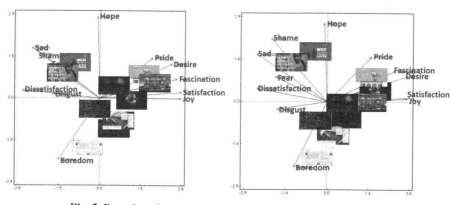

Fig. 2. Paper based   **Fig. 3.** Web based

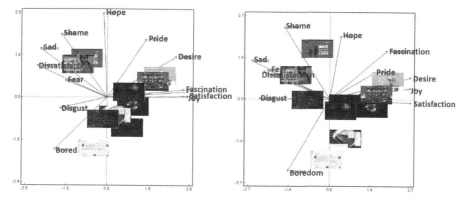

Fig. 4. Ethnic1 – paper based   **Fig. 5.** Ethnic1 – Web based

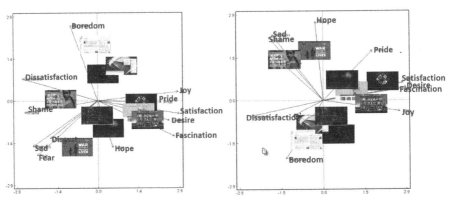

Fig. 6. Ethnic2 – paper based   **Fig. 7.** Ethnic2 – Web based

Evident in Fig. 4 and Fig. 5, the structure of Kansei that form from both measurement methods for Ethnic1 are similar. It can be observed that Hope and Boredom are also at the opposite space of both PCA plots. At far left of the plots locate Dissatisfaction, Sad and Disgust, and at most right of the plot resides Satisfaction, joy and Desire. The distributions of their relations to specimens are also quite similar from both plots. The results from these plots suggest that the use of Web-based Kansei Measurement for Ethnic1 is successful and produced desired result. It also indicates that other than simplifying the measurement process and control, the Web-based Kansei Measurement System could be used to replace the use of paper-based and Web-based Kansei measurement.

The structure of Kansei that form from Fig. 6 and Fig. 7, suggest differently. Although it can be observed that Hope and Boredom are at the opposite space of both PCA plots, the structure is unintelligible. At far left of the plots locate Dissatisfaction, and at most right of the plot resides Satisfaction, Fascination and Desire. Similarities of the distribution of their relations to specimens are unclear. However, observations made to Fig. 3, 5 and 7 shows consistency of the structure that form for both Kansei and specimen. The structure shown on Fig. 2, 4 and 6 on the other hand shows inconsistency of the structures.

The results indicate that cross-cultural Kansei measurement using traditional paper-based method will produce inconsistent results. On the contrary, measurement of cross-cultural Kansei using Web-based system will produce consistent result.

5   Conclusion

This paper investigates the feasibility of using Web-based Kansei Measurement System to measure cross-cultural Kansei in KE. KE has a limitation during measurement of Kansei, due to the use of paper based self-reporting tool, which results unique Kansei of consumers from different cultural races and demographical background [2]. The traditional measurement experiments will also take a long time to complete and the measurement experiments can only be done in local environment.

The use of Web-based Kansei Measurement System   could extend the Kansei measurement to reach larger population of consumers and of cross-cultural group of consumers. The measurement would be able to take advantage of the internet to reach multiple groups of respondent that are separated by demographical boundaries around the world.

In order to determine feasibility of Web-based Kansei Measurement System into KE, a full cycle of KE methodology in product design that uses Web-based Kansei Measurement System as Kansei measurement tool need to be investigated. This research limits itself to the initial investigation of the feasibility of Web-based Kansei Measurement System as measurement instrument and the comparative analysis was performed to observe differences and similarities of Kansei responses between Web-based Kansei Measurement System and paper based Kansei checklist. The success of Web-based Kansei Measurement System as a cross-cultural Kansei measurement instrument was also analysed by comparing the structure of Kansei by Web-based Kansei Measurement System and paper based Kansei measurement between Ethnic1 and Ethnic2. Both of the comparative studies provide encouraging results and suggest the feasibility of the integration of Web-based Kansei Measurement System into KE methodology to measure cross-cultural Kansei.

Thus, from the findings, it lends hypothetical credence and supports the hypothesis that Web-based Kansei Measurement System can be used as one of the measurement tool in Kansei Engineering methodology. However, although the Web-based Kansei Measurement System shows the positive result, the argument in Kansei Engineering that Kansei is unique with different cultural background [22][33] needs to be investigated.

Nevertheless, this research makes practical contribution to KE by providing empirical evidence that Web-based Kansei Measurement System could become potential solution in overcoming the problem of measuring global consumer's Kansei as it can be easily accessed through the internet. The feasibility of integrating Web-based Kansei Measurement System would make the cross-cultural measurement of Kansei in Kansei Engineering possible. It offers computerized data input system that would support paperless experiment environment and thus saving resources, time and energy.

Acknowledgment. This research is supported by Research Management Centre, Universiti Teknologi MARA, Malaysia [Project Code: 600-RMI/DANA 5/3/REI (6/2013)].

References

1. De Lucia, A., Francese, R., Passero, I., Tortora, G.: Slmeeting: supporting collaborative work in second life. ACM Press, New York (2008)
2. Lokman, A.M., Ishak, K.K., Razak, F.H.A., Aziz, A.A.: The feasibility of PrEmo in cross-cultural Kansei measurement. In: IEEE Symposium on Humanities, Science and Engineering Research (SHUSER). IEEE (2012)
3. Dormann, C., Chisalita, C.: Cultural values in web site design. Paper presented at the 11th European Conference on Cognitive Ergonomics (2012)

4. Bourges-Waldegg, P., Scrivener, S.A.R.: Meaning, the central issue in cross-cultural HCI design. Interacting with Computers 9(3), 287–309 (1998)
5. Suh, T., Kwon, I.W.G.: Globalization and reluctant buyers. International Marketing Review 19(6), 663–680 (2002)
6. Yoon, C., Cole, C.A., Lee, M.: Consumer Decision Making and Aging: Current Knowledge and Future Directions. Invited target article for Research Dialogues. Journal of Consumer Psychology 19(1), 2–16 (2009)
7. Moors, A.: Theories of emotion causation: A review. Cognition & Emotion 23(4), 625–662 (2009)
8. Desmet, P.M.A.: Measuring emotions. Funology: From Usability to Enjoyment. Kluwer, Dordrecht (2003)
9. Russell, J.A.: Culture and the categorization of emotions. Psychological Bulletin 110(3), 426–450 (1991)
10. Eid, M., Diener, E.: Norms for experiencing emotions in different cultures: Inter-and intranational differences. In: Culture and Well-Being, pp. 169–202 (2009)
11. Mesquita, B., Frijda, N.H., Scherer, K.R.: Culture and emotion. In: Handbook of Cross-Cultural Psychology. Basic Processes and Human Development, vol. 2, pp. 254–297 (1997)
12. Olie, R.: The 'culture' factor in personnel and organization policies'. In: International Human Resource Management, pp. 124–143. Sage Publications, London (1995)
13. Hofstede, G.: Culture's Consequences: Comparing Values, Behaviors, Institutions, and Organization Across Nations. Sage, London (2001)
14. Hofstede, G.: The Cultural Relativity of Organizational Practices and Theories. Journal of International Business Studies, 75–89 (1983)
15. Mccoy, S., Galletta, D.F., King, W.R.: Integrating National Culture into individual IS Adoption Research: The Need For Individual Level Measures. Communications of the AIS 15(12) (2005)
16. Picard, R.W., Daily, S.B.: Evaluating affective interactions: Alternatives to asking what users feel (2005)
17. Mauss, I.B., Robinson, M.D.: Measures of emotion: A review. Cognition & Emotion 23(2), 209–237 (2009)
18. Boehner, K., DePaula, R., Dourish, P., Sengers, P.: How emotion is made and measured. International Journal of Human-Computer Studies 65(4), 275–291 (2007)
19. Norman, D.A.: Measuring Emotion. The Design Journal 6(2) (2003)
20. Norman, D.A.: Emotion & design: attractive things work better. Interactions 9(4), 42 (2002)
21. Desmet, P.M.A.: Three Levels of Product Emotion. Paper presented at the International Conference on KE and Emotion Research (2010)
22. Nagamachi, M., Lokman, A.M.: Kansei Innovation: Practical Design Applications for Product and Service Development. Taylor & Francis Group (2015)
23. Lokman, A.M., Aziz, A.A.: A Kansei system to support children's clothing design in Malaysia. In: IEEE International Conference on Systems Man and Cybernetics (SMC), pp. 3669–3676 (2010)
24. Merriam-Webster Online Dictionary. Engineering (2009), http://www.merriam-webster.com/dictionary/engineering (retrieved July 27, 2009)
25. Bouchard, C., Lim, D., Aoussat, A.: Development of a KE SYSTEM for Industrial design: Identification of input data for KES (2003)
26. Robins, D., Holmes, J.: Aesthetics and credibility in web site design. Information Processing & Management 44(1), 386–399 (2008)

27. Lokman, A.M., Noor, N.M., Nagamachi, M.: ExpertKanseiWeb – A tool to design kansei website. In: Filipe, J., Cordeiro, J (eds.) ICEIS 2009. LNBIP, vol. 24, pp. 894--905. Springer, Heidelberg (2009)

28. Justis, R.T., Kreigsmann, B.: The feasibility study as a tool for venture analysis. Business Journal of Small Business Management 17(1), 35--42 (1979)

29. Georgakellos, D.A., Marcis, A.M.: Application of the semantic learning approach in the feasibility studies preparation training process. Information Systems Management 26(3) 231--240 (2009)

30. Young, G.I.M.: Feasibility studies. Appraisal Journal 38(3), 376--383 (1970)

31. Lokman, A.M.: Emotional user experience in web design: The kansei engineering approach. Universiti Teknologi MARA (2009)

32. Kaplan, B., Duchon, D.: Combining qualitative and quantitative methods in information systems research: a case study. MIS Quarterly 12(4), 571--586 (1998)

33. Ishihara, I., Nishino, T., Matsubara, Y., Tsuchiya, T., Kanda, F., Inoue, K.: Kansei and Product Development. In: M. Nagamachi (eds.), vol. 1. Kaibundo, Tokyo (2005) (in Japanese)

Part VI
Hybrid Methods

Accuracy Assessment of Urban Growth Pattern Classification Methods Using Confusion Matrix and ROC Analysis

Nur Laila Ab Ghani[1], Siti Zaleha Zainal Abidin[2], and Noor Elaiza Abd Khalid[3]

[1] College of Computer Science & Information Technology, Universiti Tenaga Nasional,
Jalan IKRAM-UNITEN, 43000 Kajang, Selangor, Malaysia
Laila@uniten.edu.my
[2,3] Faculty of Computer and Mathematical Sciences, Universiti Teknologi MARA, 40450,
Shah Alam, Selangor, Malaysia
sitizaleha533@salam.uitm.edu.my, elaiza@tmsk.uitm.edu.my

Abstract. Urban growth pattern can be categorized as either infill, expansion or outlying. Studies on urban growth classification are focusing on the description of urban growth pattern geometric features using conventional landscape metrics. These metrics are too simple and unable to give detailed information on accuracy of the classification methods. This paper aims to assess the accuracy of classification methods that can determine urban growth patterns correctly for a specific growth area. Accuracy assessments are carried out using three different classification methods – moving window, topological relation border length and landscape expansion index. Based on confusion matrices and receiver operating characteristic (ROC) analysis, results show that landscape expansion index has the best accuracy among all.

Keywords: Urban growth, pattern classification, landscape expansion index, accuracy assessment, confusion matrix, ROC analysis.

1 Introduction

Urban growth can be defined as the intensive use of urban lands and its main driving force is to accommodate the rapid increase in urban population. While urban growth is highly necessary, an unplanned and uncontrolled urban growth may lead to urban sprawl. Urban sprawl is the common term used by town planners to describe uneven pattern of the growth. However, urban sprawl classification is deemed difficult due to its unclear definition which causes conflicting views in determining sprawl patterns [1]. Inadequate information on the characteristics of urban sprawl especially in developing countries also complicates urban sprawl classification [2]. Thus, research should focus on creating an urban growth classification model instead of urban sprawl classification model for a better understanding of urban sprawl.

Urban growth patterns can be classified into three types; infill, expansion and outlying [1]. Infill refers to the development of new urban area that is mostly

© Springer Science+Business Media Singapore 2015
M.W. Berry et al. (Eds.): SCDS 2015, CCIS 545, pp. 255–264, 2015.
DOI: 10.1007/978-981-287-936-3_24

surrounded by old urban areas and usually occurs at the vacant space within old urban areas. Expansion refers to the development of a new urban area that is partially surrounded by old urban areas and usually occurs at the fringe areas. Outlying refers to the development of new separate urban areas that is not surrounded by any of old urban areas.

The patterns of urban growth can be modelled by using remote sensing data. Satellite remote sensing is a powerful tool in urban growth analysis due to its ability to produce high-quality data and updated information on the earth's surface [3]. The availability of multi-temporal datasets also provides the possible changes that occur over time before classifying urban growth patterns [4]. Based on literatures, urban growth pattern classification methods that have been used are moving window [1], topological relation border length [5] and landscape expansion index [6].

For urban growth analysis, literatures are focusing on conventional landscape metrics to describe the characteristics of an urban growth pattern. The most commonly used landscape metrics includes number of patches, percentage of landscape, edge density, landscape shape index, mean Euclidean nearest neighbor index, aggregation index and many more [7]. However, these metrics are generally focused on a simple analysis and description of geometric features [6] rather than assessing the accuracy of urban growth pattern classification methods.

This paper takes a different approach by focusing on the accuracy assessment of classification methods. For the case of changes in a map in general, the assessment is done by comparing it with a ground truth dataset obtained from the actual map. However, for the case of changes in map in specific growth patterns, the actual map does not give information about the growth patterns. This particular case requires comparison between the classification result and ground truth dataset determined by human observer who has prior knowledge about the classification criteria [8, 9, 10]. However, depending on calculating fraction of cases alone as a parameter of accuracy assessment is a bad measure that can lead to negative consequences. [11, 12]. This work proposes the use of five parameters for confusion matrix that include true positive rate, true negative rate, false positive rate, false negative rate and accuracy. Based on these parameters, the accuracy of classification methods are then presented by using receiver operating characteristic (ROC) analysis.

2   Related Works

2.1   Urban Growth Pattern Classification Methods

There are common methods for determining urban growth classifications; moving window, topological relation border length, and landscape expansion index. Moving window method uses moving window analysis and a set of classification rule to determine urban growth patterns. In the analysis, a moving window will traverse through each new urban pixel in urban growth map and the percentage of old urban pixels surrounding the pixel of interest inside the moving window is calculated. The window size can influence the classification results. To date, the sizes of moving window that ever been used in urban growth studies are 3 by 3 [13, 14] and 5 by 5 [1, 15, 16].

Topological relation border length method uses common boundary analysis and a set of classification rule to determine the urban growth patterns. In common boundary analysis, the ratio of common boundary between the old and new urban region is calculated. Infill growth occurs when the ratio is at least 0.5, expansion growth occurs when the ratio is less than 0.5 and outlying growth occurs when the ratio is equals to 0 [5, 17, 18, 19, 20].

Landscape expansion index method uses buffer zone analysis and a set of classification rule to determine the urban growth patterns. In this analysis, a buffer zone is created around the new urban region with percentage of old urban areas surrounding this new urban region (inside the buffer zone) is calculated. The buffer distance must be set roughly equal to smaller than the remote sensing data spatial resolution [6] for getting accurate values. Infill growth occurs when new urban region is surrounded by at least 50 percent of old urban areas, while expansion growth occurs when this new urban region is surrounded by less than 50 percent of old urban areas. The outlying growth occurs when new urban region is surrounded by zero percent of old urban areas [6, 21, 22].

2.2   Accuracy Assessment

Confusion Matrix

Confusion matrix is commonly used to solve a two-class or multiclass classification problem. Two-class classification involves the task of classifying cases into two set of classes with four possible combinations of either true positive, true negative, false positive or false negative. True positive is the number of actual positive cases while true negative is the number of negative cases. In contrary, false positive is the number of negative cases that are incorrectly classified as positive, and false negative is the number of positive cases that are incorrectly classified as negative [11].

Multiclass classification is an extension of two-class classification. Using the values obtained from the confusion matrix, five accuracy assessment parameters can be defined as true positive rate, true negative rate, false positive rate, false negative rate and accuracy. True positive rate or *sensitivity* represents the fraction of correctly classified positive cases. True negative rate or *specificity* represents the fraction of correctly classified negative cases. False positive rate represents the fraction of actual negative cases that are incorrectly classified, while false negative rate is the vice versa.

Table 1 summarizes the formulas for calculating such parameters based on the definitions in [11], [12], [23], [24], [25] and [26]. A good classification method must be *sensitive*, *specific* and *accurate* as much as possible [12].

Table 1. Formulas for accuracy assessment parameters.

	Two-class Model	Multiclass Model
Sensitivity	$\dfrac{TP}{TP + FN} \times 100$	$\dfrac{\sum_{i=1}^{J} TP_i}{\sum_{i=1}^{J} TP_i + \sum_{i=1}^{J} FN_i} \times 100$
Specificity	$\dfrac{TN}{TN + FP} \times 100$	$\dfrac{\sum_{i=1}^{J} TN_i}{\sum_{i=1}^{J} TN_i + \sum_{i=1}^{J} FP_i} \times 100$
False positive rate	$\dfrac{FP}{TN + FP} \times 100$	$\dfrac{\sum_{i=1}^{J} FP_i}{\sum_{i=1}^{J} TN_i + \sum_{i=1}^{J} FP_i} \times 100$
False negative rate	$\dfrac{FN}{TP + FN} \times 100$	$\dfrac{\sum_{i=1}^{J} FN_i}{\sum_{i=1}^{J} TP_i + \sum_{i=1}^{J} FN_i} \times 100$
Accuracy	$\dfrac{TP + TN}{TP + TN + FP + FN} \times 100$	$\dfrac{\sum_{i=1}^{J} TP_i + \sum_{i=1}^{J} TN_i}{\sum_{i=1}^{J} TP_i + \sum_{i=1}^{J} TN_i + \sum_{i=1}^{J} FP_i + \sum_{i=1}^{J} FN_i} \times 100$

ROC Analysis

Receiver operating characteristic (ROC) analysis is very useful in organizing, visualizing and selecting classifiers based on their performance [12, 24, 27]. The performance of classification method can be visualized using ROC graph which is a two-dimensional graph that plots the true positive rate on the y-axis and false positive rate on the x-axis. The graph represents the trade-off between benefits (true positives) and costs (false positives). A classification method that only produces a one class label or one pair of true positive and false positive rate values is called discrete classifier and it is plotted as a single point in the ROC space [24].

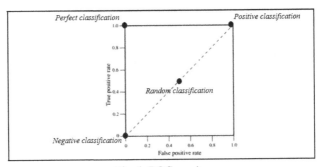

Fig. 1. ROC graph

Fig. 1 demonstrates several points in the ROC space that are important to note. The lower left point (0, 0) represents the strategy of unconditionally issuing a negative classification. In this strategy, all negative cases are correctly classified but all positive cases are also incorrectly classified as negative cases. Opposite from the lower left point strategy, the upper right point (1, 1) represents the strategy of unconditionally issuing a positive classification. In this strategy, all positive cases are correctly classified but all negative cases are also incorrectly classified as positive cases. The point (0, 1) represents perfect classification where both negative and positive cases are correctly classified. Any point along the diagonal line represents the strategy of randomly guessing a class. In this strategy, the false positive rate is equal to true positive rate. The diagonal line served as the reference line to evaluate the performance of classification technique. Any point above the diagonal line represents good classification result while any point below the diagonal line represents poor classification result. Any point nearer to the perfect classification point has better classification results than other points [12, 24].

3 Data and Methods

3.1 Data Collection

The study area is the region of Klang Valley with covering land area of latitude 2° 54' south to latitude 3° 10' south and longitude 100° 30' east to longitude 101° 47' east which is equal to 900 km^2 (30 km x 30 km). It comprises of Kuala Lumpur area, its suburbs as well as adjoining cities and towns in Selangor. This area is chosen because it is one of the most rapid urban growth areas in Malaysia. The satellite image datasets are obtained from the Department of Survey and Mapping Malaysia (JUPEM). Six Landsat Thematic Mapper (TM) images at spatial resolution of 30 meters are used where each pixel in the image represents the area of 900 square meters on the ground (1:100000). All images are acquired on 1988, 1994, 1996, 2000, 2001 and 2003 with the temporal resolution of 1 year, 2 years, 4 years and 6 years between each dataset.

3.2 Data Pre-processing

The satellite images are pre-processed into developed and undeveloped areas in Environment for Visualizing Images (ENVI) software. The process is carried out by using supervised classification which requires the user to select training sets as the basis for classification. The training sets are selected by defining two classes of region of interest (ROI); developed land and undeveloped land. Using ROIs that has been created, maximum likelihood comparison technique that will calculates and determines the class of each pixel is performed to the image.

The resulting image is saved in binary format representing only developed and undeveloped cells. Each image is in the form of bitmap image with the size of 827 pixels width and 467 pixels height, containing 386209 (827 x 467) data of both developed and undeveloped areas. Each pixel represents a ground area of 900 square meters (1:100000). The developed area is denoted as white colour with pixel value 1 while black colour represents the undeveloped area with pixel value 0. Then, image correction procedure is performed to the binary images obtained from ENVI by checking for any classification error in the images. Classification error here means areas that are already developed in preceding year are found to be undeveloped in the succeeding year. The corrected binary images are used for urban growth patterns classification.

3.3   Urban Growth Patterns Classification

The patterns of urban growth are classified using three different methods – moving window, topological relation border length and landscape expansion index. The classification is implemented using image processing toolbox in MATLAB software.

For moving window, new urban pixels are identified by comparing two binary images and the percentage of old urban pixels surrounding each new urban pixel inside the moving window is calculated. Two sizes of moving window, 3 by 3 and 5 by 5 are tested. In addition, for topological relation border length, new urban regions are identified by comparing two binary images and the ratio of common boundary between old and each new urban region is calculated. Based on the calculation, the urban growth pattern is determined using the classification rule defined by [5], [17], [18], [19] and [20]. Moreover, in landscape expansion index, new urban regions are identified by comparing two binary images and the ratio and percentage of old urban pixels surrounding each new urban region inside the buffer zone is calculated. The buffer distance is set equal to the satellite image spatial resolution which is 30 meters. Based on the calculation, the urban growth pattern is determined by using the classification rule defined by [6], [21] and [22].

In order to produce ground truth datasets, human observer must have knowledge about the characteristics of each urban growth pattern. In addition, the three unique patterns of urban growth characteristic; infill, expansion, and outlying, are considered in the pattern classification.

3.4   Accuracy Assessment

The urban growth pattern classification results are compared with ground truth dataset and the comparison results are stored in confusion matrices. A multiclass confusion matrix model is used since this research involved three patterns of urban growth. The true positive (TP), false negative (FN), true negative (TN) and false positive (FP) portion of each class in the confusion matrix can be generally summarized as follows:

$$TP = \sum_{i=1}^{3} C_{ii} \tag{1}$$

$$FN = \sum_{i=1}^{3} \sum_{j=1}^{3} C_{ij} \; ; i \neq j \tag{2}$$

$$FP = \sum_{j=1}^{3} \sum_{i=1}^{3} C_{ji} \; ; j \neq i \tag{3}$$

$$TN = \sum_{i=1}^{3} \left(N - \sum_{j=1}^{3} C_{ij} - \sum_{j=1}^{3} C_{ji} + C_{ii} \right) \tag{4}$$

where C_{ij} is the number of cases with true class i that are classified into class j, C_{ji} is the number of cases with true class j that are classified into class i, C_{ii} is the number of cases with true class i that are classified into class i and N (the total of cases in the

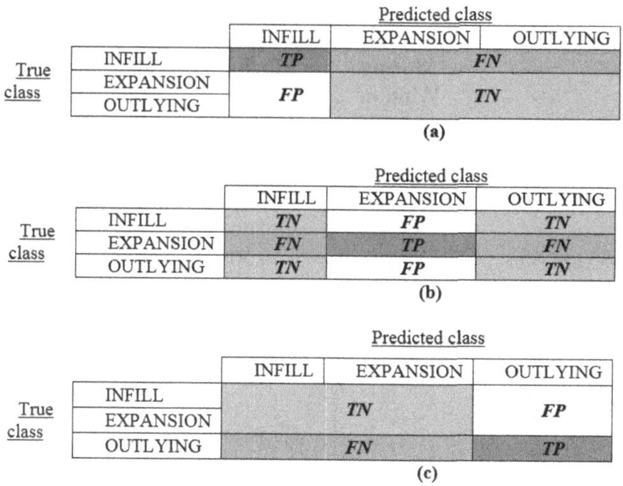

		Predicted class		
		INFILL	EXPANSION	OUTLYING
True class	INFILL	*TP*	*FN*	
	EXPANSION	*FP*	*TN*	
	OUTLYING			

(a)

		Predicted class		
		INFILL	EXPANSION	OUTLYING
True class	INFILL	*TN*	*FP*	*TN*
	EXPANSION	*FN*	*TP*	*FN*
	OUTLYING	*TN*	*FP*	*TN*

(b)

		Predicted class		
		INFILL	EXPANSION	OUTLYING
True class	INFILL	*TN*		*FP*
	EXPANSION			
	OUTLYING	*FN*		*TP*

(c)

Fig. 2. True Positives (TP), False Negative, True Negative (TN) and False Negative (FN) for (a) Infill, (b) Expansion, (c) Outlying

confusion matrix). Fig. 2 depicts these formulas for each class in the urban growth pattern model.

Five accuracy assessment parameters which are sensitivity, specificity, false positive rate, false negative rate and accuracy are calculated based on the values from confusion matrices. After the results of all calculated classification methods are obtained, ROC graph is plotted by using their respective sensitivity and false positive rate values.

4   Results and Discussion

Table 2 shows the sample results of accuracy assessment for year 1988 until 1994. Results and analyses from all datasets identify that topological relation border length and landscape expansion index produce better results when compared to moving window method. For further confirmation in finding the best method, ROC graph is used to plot the sensitivity on the y-axis and the false positive rate on the x-axis. The graph indicates better classification if its intersection point is located above the diagonal line in the ROC space and nearer to the perfect classification point (1, 1).

Fig.3(a)-3(e) show the ROC graphs for year 1988 to 2003. Four classification methods plotted in the graphs are moving window (3 by 3), moving window (5 by 5), topological relation border length (TRBL) and landscape expansion index (LEI).

From the graphs, if the conclusion is derived from the second and fifth graphs, topological relation border length method performs better than the rest. However, this method performs just slightly better than landscape expansion index method in the second graph. On the other hand, different conclusion can be made if the concluding

Table 2. Accuracy assessments for year 1988 until 1994.

	Moving Window (3 by 3)	Moving Window (5 by 5)	Topological Relation Border Length	Landscape Expansion Index
Sensitivity (%)	11.53	18.98	20.05	39.46
False negative rate (%)	88.46	81.01	79.94	60.53
Specificity (%)	55.76	59.49	60.02	69.73
False positive rate (%)	44.23	40.50	39.97	30.26
Accuracy (%)	41.02	45.99	46.70	59.64

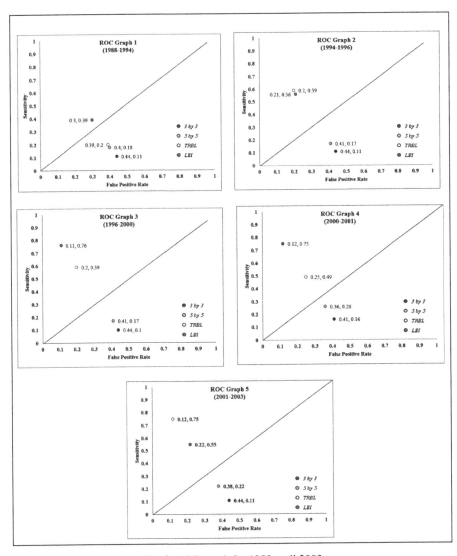

Fig. 3. ROC graph for 1988 until 2003

remarks are derived from the first, third and fourth graphs whereby the landscape expansion index method has the best performance. This remark is further strengthened by the indication that its intersection points have never located below the diagonal line in the ROC graphs when compared to topological relation border length method that is once located below the diagonal line in the first ROC graph.

5   Conclusion

This paper focuses on the accuracy assessment of three urban growth classification methods; *moving window*, *topological relation border length* and *landscape expansion index*. Based on confusion matrices and ROC graphs, we can conclude that *landscape expansion index* method performs better than the others. If the temporal resolution is included in the decision making process, we may say that the *topological relation border length* method is suitable for dataset with temporal resolution of two years duration while landscape expansion index method is suitable for dataset with temporal resolution of six years or five years or one year duration. However, this conclusion needs to be further investigated for conclusive evidence and may require additional data for larger land coverage and longer duration.

References

1. Hoffhine Wilson, E., Hurd, J.D., Civco, D.L., Prisloe, M.P., Arnold, C.: Development of a geospatial model to quantify, describe and map urban growth. Remote Sensing of Environment 86(3), 275–285 (2003)
2. Bhatta, B.: Urban Growth and Sprawl. In: Analysis of Urban Growth and Sprawl from Remote Sensing Data, pp. 1–16. Springer (2010)
3. Bagan, H., Yamagata, Y.: Landsat analysis of urban growth: How Tokyo became the world's largest megacity during the last 40years. Remote Sensing of Environment 127, 210–222 (2012)
4. Zanganeh Shahraki, S., Sauri, D., Serra, P., Modugno, S., Seifolddini, F., Pourahmad, A.: Urban sprawl pattern and land-use change detection in Yazd, Iran. Habitat International 35(4), 521–528 (2011)
5. Xu, C., Liu, M., Zhang, C., An, S., Yu, W., Chen, J.M.: The spatiotemporal dynamics of rapid urban growth in the Nanjing metropolitan region of China. Landscape Ecology 22(6), 925–937 (2007)
6. Liu, X., Li, X., Chen, Y., Tan, Z., Li, S., Ai, B.: A new landscape index for quantifying urban expansion using multi-temporal remotely sensed data. Landscape Ecology 25(5), 671–682 (2010)
7. Abebe, G.A.: Quantifying urban growth pattern in developing countries using remote sensing and spatial metrics: A case study in Kampala, Uganda (Doctoral dissertation, MS thesis submitted to the Faculty of Geo-Information Science and Earth Observation of the University of Twente) (2013)
8. Tagashira, H., Arakawa, K., Yoshimoto, M., Mochizuki, T., Murase, K., Yoshida, H.: Improvement of lung abnormality detection in computed radiography using multi-objective frequency processing: Evaluation by receiver operating characteristics (ROC) analysis. European Journal of Radiology 65(3), 473–477 (2008)

9. Barnes, M., Duckett, T., Cielniak, G., Stroud, G., Harper, G.: Visual detection of blemishes in potatoes using minimalist boosted classifiers. Journal of Food Engineering 98(3), 339–346 (2010)
10. Xu, L., Li, J., Brenning, A.: A comparative study of different classification techniques for marine oil spill identification using RADARSAT-1 imagery. Remote Sensing of Environment 141, 14–23 (2014)
11. Dziuda, D.M.: Data mining for genomics and proteomics: analysis of gene and protein expression data, vol. 1. John Wiley & Sons (2010)
12. Gorunescu, F.: Data Mining: Concepts, models and techniques, vol. 12. Springer (2011)
13. Kumar, U., Mukhopadhyay, C., Ramachandra, T.V.: Spatial Data Mining and Modeling for Visualisation of Rapid Urbanisation. SCIT Journal 9 (2009)
14. Abiden, M.Z.Z., Abidin, S.Z., Jamaluddin, M.N.F.: Pixel based urban growth model for predicting future pattern. In: 2010 6th International Colloquium on Signal Processing and Its Applications (CSPA), pp. 1–5. IEEE (2010)
15. Pham, H.M., Yamaguchi, Y., Bui, T.Q.: A case study on the relation between city planning and urban growth using remote sensing and spatial metrics. Landscape and Urban Planning 100(3), 223–230 (2011)
16. Ab Ghani, N.L., Abidin, S.Z., Abiden, M.Z.Z.: Generating Transition Rules of Cellular Automata for Urban Growth Prediction. International Journal of Geology 5(2), 41–47 (2011)
17. García, A.M., Santé, I., Boullón, M., Crecente, R.: A comparative analysis of cellular automata models for simulation of small urban areas in Galicia, NW Spain. Computers, Environment and Urban Systems 36(4), 291–301 (2012)
18. Hao, R., Su, W., Yu, D.: Quantifying the Type of Urban Sprawl and Dynamic Changes in Shenzhen. In: Li, D., Chen, Y. (eds.) Computer and Computing Technologies in Agriculture VI, Part II. IFIP AICT, vol. 393, pp. 407–415. Springer, Heidelberg (2013)
19. Sun, C., Wu, Z.F., Lv, Z.Q., Yao, N., Wei, J.B.: Quantifying different types of urban growth and the change dynamic in Guangzhou using multi-temporal remote sensing data. International Journal of Applied Earth Observation and Geoinformation 21, 409–417 (2013)
20. Yue, W., Liu, Y., Fan, P.: Measuring urban sprawl and its drivers in large Chinese cities: The case of Hangzhou. Land Use Policy 31, 358–370 (2013)
21. Zeng, Y., Xu, Y., Li, S., He, L., Yu, F., Zhen, Z., Cai, C.: Quantitative analysis of urban expansion in central china. International Archives of the Photogrammetry, Remote Sensing and Spatial Information Sciences 39(B7), 363–366 (2012)
22. Li, C., Li, J., Wu, J.: Quantifying the speed, growth modes, and landscape pattern changes of urbanization: a hierarchical patch dynamics approach. Landscape Ecology 28(10), 1875–1888 (2013)
23. Gupta, G.K.: Introduction to data mining with case studies. PHI Learning Pvt. Ltd. (2006)
24. Fawcett, T.: An introduction to ROC analysis. Pattern Recognition Letters 27(8), 861–874 (2006)
25. Shi, X., Cheng, H.D., Hu, L., Ju, W., Tian, J.: Detection and classification of masses in breast ultrasound images. Digital Signal Processing 20(3), 824–836 (2010)
26. Sokolova, M., Lapalme, G.: A systematic analysis of performance measures for classification tasks. Information Processing & Management 45(4), 427–437 (2009)
27. Metz, C.E.: Basic principles of ROC analysis. In: Seminars in Nuclear Medicine, vol. 8(4), pp. 283–298. WB Saunders (October 1978)

Intrusion Detection System Based on Modified K-means and Multi-level Support Vector Machines

Wathiq Laftah Al-Yaseen[1,2], Zulaiha Ali Othman[1], and Mohd Zakree Ahmad Nazri[1]

[1] Data Mining and Optimization Research Group (DMO)
Centre for Artificial Intelligence Technology (CAIT)
School of Computer Science, Faculty of Information Science and Technology
Universiti Kebangsaan Malaysia (UKM), 43600 Bandar Baru Bangi, Malaysia
[2] Al-Furat Al-Awsat Technical University
wathiqpro@gmail.com, {zao,zakree}@ukm.edu.my

Abstract. This paper proposed a multi-level model for intrusion detection that combines the two techniques of modified K-means and support vector machine (SVM). Modified K-means is used to reduce the number of instances in a training data set and to construct new training data sets with high-quality instances. The new, high-quality training data sets are then utilized to train SVM classifiers. Consequently, the multi-level SVMs are employed to classify the testing data sets with high performance. The well-known KDD Cup 1999 data set is used to evaluate the proposed system; 10% KDD is applied for training, and corrected KDD is utilized intesting. The experiments demonstrate that the proposed model effectively detects attacks in the DoS, R2L, and U2R categories. It also exhibits a maximum overall accuracy of 95.71%.

Keywords: intrusion detection system, network security, support vector machine, K-means, multi-level SVM.

1 Introduction

Intrusion detection systems (IDS) limit the serious influence of attacks on system resources. They are used as tools behind firewalls to identify suspicious patterns by monitoring and analyzing the events in a computer network. IDS is classified as either a signature or an anomaly detection system [1]. Signature detection systems (misuse detection systems) aim to determine the defined patterns or signatures of attacks in traffic networks. These systems identify known attacks efficiently but fail to detect new attacks (zero-day attacks) whose signatures have not been saved previously in the database. By contrast, anomaly detection systems identify new attacks by learning the normal behavior of the system and then generating an alarm in the event of a deviation from the normal behavior. This deviation is considered an intrusion [2].Therefore, anomaly detection systems report higher false alarm rates than signature detection systems do.

In many approaches, anomaly detection systems are implemented with different techniques to improve IDS accuracy, as discussed in the subsequent section. A popular technique used with IDS is the support vector machine (SVM). This technique has

© Springer Science+Business Media Singapore 2015
M.W. Berry et al. (Eds.): SCDS 2015, CCIS 545, pp. 265–274, 2015.
DOI: 10.1007/978-981-287-936-3_25

satisfactorily classified data, particularly in conjunction with IDS. Nonetheless, these results rely heavily on the quality of the training data set used to train SVMs. If the training data set is large, then the training complexity of SVM is high. This occurrence may cause system failure because of the high consumption of memory [3]. Given that the majority of training data sets for IDS is large, including the KDD Cup 1999 data set, these defects must be addressed by reducing the number of instances in training data sets. Some researchers have removed redundant instances from data sets as a preprocessing step as in [4, 5], whereas others have used techniques to reduce the size of training data sets, as in [3].

In the present study, we propose a model that utilizes a modified K-means algorithm at the preprocessing stage to reduce the number of instances for training data sets. This model also uses SVM as a multi-level classifier to build an anomaly intrusion detection system that can detect unknown attacks. We select the K-means algorithm because of its capability to cluster instances into highly similar groups. We employ this algorithm to generate a new training data set that represents all instances in the original training data set by improving the method of selecting the initial centroids of clusters that represent all cases. First, a training data set is separated during preprocessing into five categories: Normal, DoS, Probe, R2L, and U2R. Then, the number of instances in each category is reduced through modified K-means while maintaining the high quality of the categories for the training data set. The resultant five categories of the data set are then employed to learn multi-level SVMs. The proposed model can reduce training time and achieve a favorable detection performance as a result of IDS. The remainder of this paper is organized as follows. Section 2 provides an overview of the K-means algorithm and the SVM classifier. Section 3 describes the proposed system. Section 4 presents the experimental results. Finally, Section 5 provides the conclusion.

2   Related Work

Many machine learning and data mining techniques have recently been proposed to design IDS models that can detect known and unknown attacks. However, the detection and false alarm rates of an anomaly intrusion detection system remain poor. Some of these models combine two or more techniques to improve accuracy. In the current study, we review previous studies related to the selection of the initial centroids of clusters for K-means and the studies that use multi-levels to implement classifiers for IDS. All of the following studies employ the KDD Cup 1999 data set to evaluate performance.

The K-means algorithm is highly sensitive to the initial centroids of clusters. In fact, many studies seek to improve the method of selecting the optimal initial centroids of clusters. These centroids effectively separate clusters and accelerate their convergence behavior. The initialization methods for K-means were investigated comparatively by Celebi et al. [6]. Gao and Wang [7] identified the initial center of clusters as instances with the least similar degree of information entropy. Sujatha and Sona [8] proposed the initial method to enhance K-means, in which the sensitivity of local minima and the randomness for K-means are reduced. However, this method has a long processing time. Kathiresan and Sumathi [9] utilized the Z-score ranking method to select improved initial centroids for K-means. Nonetheless, the complexity time is long given large data sets. Nazeer and Sebastian [10] proposed an iterative process to select initial centroids in which the distance of each data point from all other data points must be calculated. Therefore, a large set of data points requires much computation.

Multi-level models were successfully used to construct IDS and to improve detection accuracy. Pfahringer [11] presented the bagged boosting of C5 as a model for IDS. Xiang et al. [12] proposed multiple tree classifier models that employ the C4.5 technique at each level. The DoS, Probe, and Normal categories were classified at the first level, whereas R2L and U2R were classified at the second level. In 2008, Xiang et al. [13] presented another multi-level hybrid classifier that combines decision trees and Bayesian clustering. The C4.5 model was used to extract DoS and Probe attacks. Then, Bayesian clustering (AutoClass technique) was employed to cluster the R2L, U2R, and Normal categories. The largest cluster represents the Normal classes, whereas the other clusters denote R2L and U2R attacks. The AdaBoost algorithm with a single weak classifier was proposed by Natesan et al. [14] to build IDS. The classifiers used in this algorithm are Bayes Net, Naïve Bayes, and decision trees. Ambwani [15] presented the multi-class SVM that uses the one-versus-one method to classify each attack. Nonetheless, the proposed method is no better than the winner method established in [11]. Ambwani [16] also presented a model that uses neural network and fuzzy theory to reduce the high rate of false positive alarms. This study analyzes the advantages and disadvantages of neural network and fuzzy logic. It then generates a new model with enhanced generalization, learning, and mapping capability. Lu and Xu [17] proposed a three-level hybrid IDS that combines supervised classifiers such as C4.5 and Naïve Bayes with unsupervised clustering (i.e., Bayesian clustering) in different levels to classify various classes. In the first level, the C4.5 algorithm was used to separate the data set into three categories: DoS, Probe, and Others. Naïve Bayes was used to distinguish the U2R category from the other categories in the second level. In the third level, Bayesian clustering separated the category R2L from Normal with high detection. Finally, Gogoi et al. [18] proposed a multi-level hybrid intrusion detection system that combines supervised, unsupervised, and outlier methods to improve detection rate. The proposed method classified the DoS and Probe categories at the first level using the CatSub+ supervised classifier. In the second level, the unsupervised classifier K-point algorithm was applied to distinguish the Normal category from the rest of the test data set. In the final level, the remaining data were grouped into R2L and U2R using the outlier-based classifier GBBK.

In summary, all studies that employ K-means for IDS attempt to improve performance by enhancing the method of selecting the initial centroids of clusters. However, these methods are flawed in terms of the increased complexity of processing time and the fact that each resultant cluster retains many instances from different classes. Moreover, choosing the best sequence with which to classify classes with high accuracy remains difficult for the multi-level classifier model.

3   Proposed Modified K-means and Multi-level SVMs Model

The proposed model that combines modified K-means with multi-level SVMs is described in this section. We thus summarize its steps as follows:

- The training data set is examined (10% KDD data set).
- The symbolic attributes *protocol type*, *service*, and *flag* are converted into numeric types, as in [19].
- The training data set is normalized to [0, 1], as demonstrated in [19].
- The training data set is divided into five categories (Normal, DoS, Probe, R2L, and U2R).
- Modified K-means is applied to each category to generate five new training data sets.
- Each SVM is trained with one of the new training data sets.
- The first three steps are repeated to test the data set (corrected KDD data set).
- The multi-level SVMs in Fig. 1 are applied to classify the instances of testing the data set.
- The performance of the model is assessed in terms of accuracy, detection rate, and measures of false alarm rate.

Before training with SVM, the training data should be preprocessed, such as by converting and normalizing attributes. Then, the training data set is divided into the Normal, DoS, Probe, R2L, and U2R categories. Modified K-means is applied to each category to reduce the number of instances by clustering and by computing the average of each cluster as a new instance. For example, the result is a set of clusters with similar instances when modified K-means is implemented in the Normal category. Thus, the instances of the new Normal category are represented by computing the average of each cluster as a new instance. Table 1 shows the number of instances of the 10% KDD Cup 1999 data set before and after this stage. The quality of the resultant instances represents that of all of the instances in the original training data set.

Table 1. Number of instances in the 10% KDD Cup 1999 data set before and after categorization and applying modified K-means.

Category	# of instances (before)	# of instances (after)
Normal	97,278	639
DoS	391,458	140
Probe	4,107	134
R2L	1,126	51
U2R	52	25
Total	494,021	989

The K-means algorithm depends on two factors, namely, the number of clusters and the initial centroids of clusters, to optimize the clustering of instances [20]. The details and pseudo-code of standard K-means are shown in [21]. Our modified K-means must specify these two factors to identify a threshold value as the maximum distance between the centroid of clusters and the instances of the data set. Algorithm 1 shows the steps of the modified K-means algorithm. The number of clusters k is computed dynamically without requiring the user's input (steps 1 and 2), unlike in the standard algorithm. The modified algorithm computes the initial centroids of clusters by searching for all of the instances in a data set with distances that are larger than the threshold, as indicated in steps 1 and 2 of Algorithm 1, whereas the standard algorithm generates these instances randomly. Accordingly, the differences between the modified and standard K-means are presented in steps 1 and 2 of Algorithm 1.

Algorithm 1. Modified K-means algorithm

Input: Whole instances of category D

Output: High quality instances of category D′

Step 1. Set k = 1, c_1 = First instance $\omega_1 \in D$

Step 2. For every instance $\omega_l \in D$ and $i \neq 1$ Do

Step 2.1. If $\|\omega_i - c_s\| > threshold, s = 1, ..., k$ Then

Step 2.2. k = k +1, $c_k = \omega_i$

Step 3. Assign every instance $\omega_l \in D$ to closest centroid in order to make k Clusters {C_1, C_2, ..., C_k}

Step 4. Calculate cluster centroids $\overline{\omega_i} = \frac{1}{k_i}\sum_{j=1}^{k_i} \omega_{ij}, i = 1, ..., k$

Step 5. For every instance $\omega_l \in D$ Do

Step 5.1. Reassign ω_i to closest cluster centroid; $\omega_i \in C_s$ is moved from C_s to C_t

If $\|\omega_i - \overline{\omega_t}\| \leq \|\omega_i - \overline{\omega_j}\| for all j = 1, ..., k, j \neq s$.

Step 5.2. Recalculate centroids for clusters C_s and C_t.

Step 6. If cluster instances are stabilized Then (D′ = centroids of clusters) Else go to Step 4.

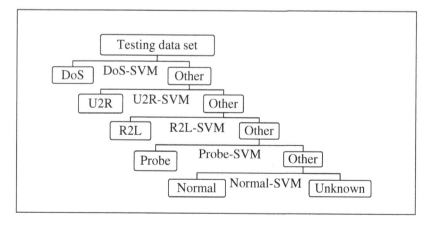

Fig. 1. Multi-level classification testing data set using SVM

Therefore, the proposed system generates five new training data sets from these categories. The first is the normal training data set, which considers normal instances as class 1 and the other instances of other categories are class 2. The same steps are repeated with the other categories. Finally, the system uses these new training data sets to learn five SVM classifiers that utilize different parameters to improve the performance of IDS. These SVMs are Normal-SVM, DoS-SVM, Probe-SVM, R2L-SVM, and U2R-SVM.

In the testing phase, the converted and normalized data are preprocessed to test the data set. Then, the multi-level classification depicted in Fig. 1 is applied. Previous studies [13, 17, 18] have proposed different multi-level methods to implement classifiers, as discussed in the section on Related Works. The ideal multi-level classification of the applied SVMs for testing a data set is derived from the results of several experiments, as exhibited in Fig. 1. DoS-SVM is implemented first because the classes of DoS have been less similarity to other categories given that this kind of intruder does not use the legitimate behavior of a user during attacks. In the subsequent levels, U2R, R2L, and Probe are classified according to the amount of instances for each category. U2R attacks involve the fewest instances in comparison with the other attacks. While the risk of these attacks is significant, U2R attacks are considered the most dangerous. In addition, the reason behind extracting these categories before normal is the similarity factor among their instances and those of the Normal category. The final SVM applied to the proposed system is Normal-SVM, which separates normal instances from the remaining instances. The remaining instances that are not classified under any category are considered unknown attacks.

4   Experimental Results

To ensure experimental persuasiveness and convenience, the proposed system uses the KDD Cup 1999 data sets as benchmarks to evaluate the experiments. These data sets originated from the Lincoln Laboratory of the Massachusetts Institute of Technology. They were developed by DARPA and are considered standard benchmarks for the evaluation of intrusion detection systems. The training and testing data sets of KDD Cup 1999 contain 4,898,431 and 311,029 instances, respectively. All instances of these data sets fall into the five main categories Normal, DoS, Probe, R2L, and U2R. The training data set contains 22 types of attacks in addition to those in the normal class, whereas the testing data set contains only an additional 17 types of attacks. Each instance in the data set displays41 continuous and discrete features [13, 22].

In this experiment, we use the 10% KDD data set for training. This data set contains 494,021 instances. The corrected KDD data set is utilized for testing and contains 311,029 instances. A computer that runs on an Intel Core i5 processor with 2.60 GHz and 12 GB RAM is employed. The freeware package LibSVM [23] is coded using Java to implement the proposed system. We apply nu-SVC and RBF kernels to run the LibSVM in this study, and the ideal values of the parameters nu and gamma are determined for each category as per the results of several experiments, as listed in Table 2. The threshold value to reduce the number of instances for all categories using modified K-means is 0.5, as indicated in Table 1. The popular measures of intrusion detection systems, such as accuracy, detection rate (recall), and false alarm rate are used to evaluate system performance.

Table 2. Parameter values of the nu-SVC classifier

Category	nu	gamma (γ)
Normal	0.06	0.09
DoS	0.004	0.5
Probe	0.1	0.3
R2L	0.05	0.008
U2R	0.05	0.008

The best performance of this system in terms of accuracy is 95.71%, that of detection rate is 95.02%, and that of false alarm rate is 1.45%. The details of the results are shown in the confusion matrix of Table 3.

The detection rates of R2L and U2R are minimal in comparison with those of DoS and Probe because the number of instances in these attacks is much less than that in DoS and Probe given the KDD Cup 1999 data set. Concurrently, two types of attacks are found in R2L:7,741 snmp get attack and 2,406 snmp guess. Their features are highly similar to those of Normal and may match these features100%. Hence, the predicate number of R2L as Normal is high. The proposed model with combined standard K-means is initially compared with multi-level SVMs to highlight the capability of the modified K-means to build a new training data set with high-quality instances. To compute the results of standard K-means, we must identify the best number of clusters. The ideal value of k is 90, at which the accuracy is high at 91.65%. Therefore, we can compare the results of the proposed model with those of the combined standard K-means and the multi-level SVMs when k is equal to 90. The accuracy, detection rate, and false alarm rate of the proposed method are generally enhanced under the combined standard K-means with multi-level SVM, with the exception of the detection rate of U2R.

Table 3. Confusion matrix for the proposed system with 10%KDD for training and corrected KDD for testing

		Predicate						Total	Re-call
		Normal	DoS	Probe	R2L	U2R	Unknown		
	Normal	59714	84	116	255	7	417	60593	98.55
	DoS	722	223347	107	148	0	5529	229853	99.57
Actual	Probe	598	193	2885	3	0	487	4166	80.94
	R2L	11060	1	1	1603	6	3518	16189	31.63
	U2R	101	0	74	16	26	11	228	16.23
	Total	72195	223625	3183	2025	39	9962	311029	

Table 4. Comparison of the proposed method and the combined standard K-means with multi-level SVMs

Method	Normal	DoS	Probe	R2L	U2R	Accuracy	FAR
Standard K-means with multi-level SVMs	88.93	96.57	69.14	6.26	**52.85**	91.65	11.07
Proposed method	**98.55**	**99.57**	**80.94**	**31.63**	16.23	**95.71**	**1.45**

Therefore, we compare the performance of the proposed model with that of other methods, such as the bagged boosted (Winner's) [11], multi-class SVM [15], neuro-fuzzy controller(NFC; artificial neural network and fuzzy) [16], adaptive importance sampling (AIS),multi-object genetic fuzzy IDS (MOGFIDS; GA and fuzzy) [24], balance iterative reducing and clustering using hierarchies (BIRCH), and SVMs [3], as depicted in Table 5.

The proposed method is the most accurate overall among the other methods and reports the best detection rates for DoS and R2Lattacks. The multi-class SVM has the best detection rate for the Normal class, whereas those for the other categories are worse than the rates obtained with other methods. Hierarchical BIRCH and SVM achieve high detection rates for attacks in Probe and U2R.Moreover, the accuracy and detection rates of Normal and DoS are moderate and are close to the results of the proposed model. Therefore, the detection rate of Probe decreases with the increase in the Normal or DoS detection rates. The reason of detection rate of Normal category is small compared with the other methods due to the level of Normal-SVM in multi-level model is the last one as depicted in Fig. 1.For instance, when change the level of Normal-SVM to the first level, then the detection rate of Normal will be increased, but this change will effect on the performance of the proposed model with the other categories like R2L and U2R. However, the detection rate of the proposed method for the attacks in Probe category is less than other methods because there is a type of new attack called MScan belong to Probe has a low detection rate with SVM classifier. Consequently, the overall detection rate of Probe with SVM is small. We believe that the proposed method generates the best results in relation to the balance state among all of the categories. As a result, its accuracy exceeds those of other methods. Several methods are employed to evaluate the proposed IDS, such as 10-fold cross validation or the application of the same data set for training and testing. Some methods also utilize data sets that are generated randomly from the original KDD Cup 1999 data set. Hence, performance is high. Consequently, any proposed method for IDS should be compared according to the same evaluation method. Thus, we use the methods in Table 5 only for comparison given that the best evaluation method for IDS involves training and testing the KDD Cup 1999 data sets.

Table 5. Comparison with other methods in terms ofdetection rate, accuracy, and false alarm rate

Method	Normal	DoS	Probe	R2L	U2R	Accuracy	FAR
Winner's (2000)	99.50	97.10	83.30	8.40	13.20	93.30	0.55
Multiclass SVM (2003)	**99.6**	96.8	75	4.2	5.3	92.46	**0.43**
MOGFIDS (2007)	98.36	97.20	88.6	11.01	15.79	93.20	1.6
BIRCH and SVM (2011)	99.3	99.5	**97.5**	28.8	**19.7**	95.7	0.7
NFC (2014)	98.2	99.5	84.1	31.5	14.1	N/A	1.9
Proposed method	98.55	**99.57**	80.94	**31.63**	16.23	**95.71**	1.45

5   Conclusion

In this paper, we proposed a model of modified K-means with multi-level SVMs to construct a high-performance intrusion detection system. Modified K-means was applied to reduce the number of training data sets and to obtain new, high-quality training data sets with which to learn SVMs. The nu-SVM and RBF kernel functions of LibSVM were employed to implement multi-level SVMs. The converted and normalized training and testing data sets were preprocessed to render them suitable for the SVM classifier. This model classified the attacks in DoS, R2L, and U2R effectively. In addition, its capability to classify other types of instances, such as Normal and Probe, is not worse than those of other models. In future studies, we attempt to improve performance in relation to the Normal and Probe categories and conduct comparisons with other studies that employ different evaluation methods.

References

1. Ghanem, T.F., Elkilani, W.S., Abdul-kader, H.M.: A hybrid approach for efficient anomaly detection using metaheuristic methods. J. Adv. Res. Article in Press (2014)
2. Om, H., Kundu, A.: A hybrid system for reducing the false alarm rate of anomaly intrusion detection system. In: 1st International Conference on Recent Advances in Information Technology (RAIT), pp. 131–136. IEEE (2012)
3. Horng, S.-J., Su, M.-Y., Chen, Y.-H., et al.: A novel intrusion detection system based on hierarchical clustering and support vector machines. Expert Syst. Appl. 38, 306–313 (2011)
4. Hasan, M., Nasser, M., Pal, B., Ahmad, S.: Intrusion Detection Using Combination of Various Kernels Based Support Vector Machine. International Journal of Scientific & Engineering Research 4, 1454–1463 (2013)
5. Yao, J., Zhao, S., Fan, L.: An enhanced support vector machine model for intrusion detection. In: Wang, G.-Y., Peters, J.F., Skowron, A., Yao, Y. (eds.) RSKT 2006. LNCS (LNAI), vol. 4062, pp. 538–543. Springer, Heidelberg (2006)
6. Celebi, M.E., Kingravi, H.A., Vela, P.A.: A comparative study of efficient initialization methods for the k-means clustering algorithm. Expert Syst. Appl. 40, 200–210 (2013)
7. Gao, M., Wang, N.: A Network Intrusion Detection Method Based on Improved K-means Algorithm. Adv. Sci. Technol. Lett. 53, 429–433 (2014)
8. Sujatha, M.S., Sona, M.A.S.: New fast k-means clustering algorithm using modified centroid selection method. International Journal of Engineering Research and Technology 2, 1–9 (2013)

9. Kathiresan, V., Sumathi, P.: An efficient clustering algorithm based on Z-Score ranking method. In: International Conference on Computer Communication and Informatics (ICCCI), pp. 1–4. IEEE (2012)
10. Nazeer, K.A., Sebastian, M.: Improving the Accuracy and Efficiency of the k-means Clustering Algorithm. In: Proceedings of the World Congress on Engineering, vol. 1, pp. 1–3 (2009)
11. Pfahringer, B.: Winning the KDD99 classification cup: bagged boosting. ACM SIGKDD Explorations Newsletter 1, 65–66 (2000)
12. Xiang, C., Chong, M., Zhu, H.: Design of mnitiple-level tree classifiers for intrusion detection system. In: 2004 IEEE Conference on Cybernetics and Intelligent Systems, vol. 2, pp. 873–878. IEEE (2004)
13. Xiang, C., Yong, P.C., Meng, L.S.: Design of multiple-level hybrid classifier for intrusion detection system using Bayesian clustering and decision trees. Pattern Recognit. Lett. 29, 918–924 (2008)
14. Natesan, P., Balasubramanie, P., Gowrison, G.: Improving the Attack Detection Rate in Network Intrusion Detection using Adaboost Algorithm. Journal of Computer Science 8, 1041–1048 (2012)
15. Ambwani, T.: Multi class support vector machine implementation to intrusion detection. In: Proceedings of the International Joint Conference on Neural Networks, vol. 3, pp. 2300–2305. IEEE (2003)
16. He, L.: An Improved Intrusion Detection based on Neural Network and Fuzzy Algorithm. Journal of Networks 9, 1274–1280 (2014)
17. Lu, H., Xu, J.: Three-level Hybrid Intrusion detection system. In: International Conference on Information Engineering and Computer Science, ICIECS 2009, pp. 1–4. IEEE (2009)
18. Gogoi, P., Bhattacharyya, D., Borah, B., Kalita, J.K.: MLH-IDS: A Multi-Level Hybrid Intrusion Detection Method. The Computer Journal 57, 602–623 (2014)
19. Sabhnani, M., Serpen, G.: Application of Machine Learning Algorithms to KDD Intrusion Detection Dataset within Misuse Detection Context. In: MLMTA, pp. 209–215 (2003)
20. Jianliang, M., Haikun, S., Ling, B.: The application on intrusion detection based on k-means cluster algorithm. In: International Forum on Information Technology and Applications, IFITA 2009, vol. 1, pp. 150–152. IEEE (2009)
21. Bhatia, M., Khurana, D.: Experimental study of Data clustering using k-Means and modified algorithms. International Journal of Data Mining & Knowledge Management Process (IJDKP) 3, 17–30 (2013)
22. KDD Cup 1999 Data set. http://archive.ics.uci.edu/ml/machine-learning-databases/kddcup99-mld/
23. LibSVM. http://www.csie.ntu.edu.tw/~cjlin/libsvm/
24. Tsang, C.-H., Kwong, S., Wang, H.: Genetic-fuzzy rule mining approach and evaluation of feature selection techniques for anomaly intrusion detection. Pattern Recognit. 40, 2373–2391 (2007)

Author Index

Printed in the United States
By Bookmasters